Acknowledgments

Although our names appear alone on the cover of this book, many people have contributed in some form or other to the book's creation. In many cases, these people are good friend of ours; and in other cases, we have never met the individuals and have conversed with them only on the phone or by email. We thank you all who helped us, as we are certain that we could not have completed this book without the help, assistance, and moral support.

We must thank Anis's wife and his children for their understanding and support while Anis was busy late nights and weekends working on the book. We also extend our thanks to Mr. A. Jalil for believing in Anis and opening a world of opportunities for him.

We thank Una Cogavin, our personal friend, who helped us edit some of the chapters at times when we were scrambling to meet the deadlines. Una provided us with feedback that helped us do a better job at writing.

Anis and I are both extremely thankful to Dr. Bob Harbort who was instrumental in our academic careers. Dr. Harbort taught us the information research process in those days when research tools like the Internet were unheard of.

We must also thank Dr. Doreen Galli Erickson, one of the best mentors on this planet, who helped us build our computer science foundation and introduced advanced computing concepts to us. We also thank Mr. Mohibullah Sheikh, the brilliant mathematician and beloved teacher, who taught us how to think critically and approach problems rationally.

Margaret Eldridge, our initial editor for this book at Wiley Publishing, deserves an award for the amount of effort and dedication she gave us. We are sure that she had no idea what she was getting into. Margaret taught us more about writing in the short time we spent with her than I learned in all my years. Margaret, thanks for giving us this opportunity. And thanks, too, to Carol Long for shepherding this project to completion during the past few months.

Scott Amerman, our development editor at John Wiley and Sons, worked incredibly hard on the manuscripts and the overall book contents. He has been absolutely indefatigable while dealing with the manuscript changes as we worked on the manuscript at the same time. We appreciate his patience and understanding in working with two very green writers.

Michelle Ragsdale and Mark Shapiro of Davis Marrin, the public relations firm of Agere Corporation, provided us with information on Agere Wireless LAN products. We are extremely thankful to them for accommodating our needs on extremely short notice.

*We dedicate this book to our parents
for their hard work and countless sacrifices,
which helped us reach where we are today.*

Contents

Acknowledgments		xiii
About the Authors		xv
Introduction		xvii
PART 1	**Introduction to Wireless Local Area Networks (LANs)**	1
Chapter 1	**Networking Basics**	3
	Development of Computer Networks: An Overview	4
	Network Types	8
	Peer-to-Peer Networks	8
	Local Area Networks (LANs)	9
	Wide Area Networks (WANs)	9
	Personal Area Networks (PANs)	11
	The Internet	11
	Virtual Private Networks (VPNs)	12
	Network Topologies	13
	Three Commonly Used Topologies	13
	Choosing the Right Topology	15
	Network Hardware and Software	17
	Networking Components	17
	Networking Software	28
	Networking Protocol: TCP/IP	28
	Putting It All Together	32
	Summary	33
Chapter 2	**Wireless LANs**	35
	Evolution of Wireless LANs: An Overview	36
	A Basic Wireless LAN	37

	Basic Architecture of a Wireless LAN	39
	Wireless LAN Adapters	40
	Access Points (APs)	47
	Wireless LAN Configurations	49
	Ad-Hoc Mode	49
	Infrastructure Mode	49
	Distribution Service Systems (DSSs)	50
	Existing Wireless LAN Standards	51
	IEEE 802.11	52
	IEEE 802.11b	52
	IEEE 802.11a	52
	HomeRF	52
	Bluetooth	53
	Are Wireless LANs Risks to Health?	53
	Security Risks	53
	Summary	54
Chapter 3	**The Institute of Electrical and Electronics Engineers (IEEE) 802.11 Standards**	**55**
	History of IEEE	56
	IEEE 802 Wireless Standards	56
	The 802.11 Working Group	57
	The 802.15 Working Group	57
	The 802.16 Working Group	58
	The 802.11 Family of Standards	58
	The 802.11 Standard Details	59
	802.11 Security	61
	Operating Modes	62
	Roaming	63
	The 802.11 Extensions	64
	802.11b	64
	802.11a	66
	802.11g	68
	802.11 Shortcomings	69
	Wireless Standards Comparison	69
	Summary	70
Chapter 4	**Is Wireless LAN Right for You?**	**71**
	Benefits of Wireless LANs	72
	Deployment Scenarios	73
	Small Office Home Office (SoHo)	73
	Enterprise	74
	Wireless Internet Service Providers (WISPs)	75
	Costs Associated with Wireless LANs	78
	SoHo	79
	Enterprise	79
	WISPs	79

		Deployment Issues	79
		SoHo	80
		Enterprise	80
		WISPs	80
	Security		81
	Health Concerns		81
	Summary		81
PART 2	**Secure Wireless LANs**		**83**
Chapter 5	**Network Security**		**85**
	Network Operational Security		86
		Physical Security	87
	Common Network Attacks on Operational Security		94
		External Network Attacks	94
		Internal Network Attacks	101
	Network Data Security		103
		Resident-Data or File Security	104
		Protecting Data Using Cryptographic Primitives	104
	Network Data Transmission and Link Security		106
		Securing Network Transmission	108
	Summary		116
Chapter 6	**Securing the IEEE 802.11 Wireless LANs**		**117**
	Wireless LAN Security Requirements		118
		Wireless LAN Operational Security Requirements	119
		Wireless LAN Data Security	122
	The Institute of Electrical and Electronics Engineers (IEEE) 802.11 Standard Security		123
		Service Set Identifiers (SSID)	123
		Wired Equivalent Privacy (WEP) Protocol	123
	IEEE 802.11 WEP Protocol Weaknesses and Shortcomings		129
	The Future of 802.11 Standard Security		131
	Common Security Oversights		131
		Using Default or Out-of-the-Box Security	131
		Using Fixed Shared Keys	132
		Using Far-Too-Strong Radio Signals	132
	Extending Wireless LAN Security		132
		The 802.1X Authentication Protocol	132
		Virtual Private Networks (VPNs)	136
	Securing Wireless LAN		137
		User Authentication	138
		Data Confidentiality and Privacy	138
		Wireless LAN Passwords and Usage Policies	139
		Frequent Network Traffic and Usage Analysis	139
	Summary		139

Contents

PART 3	**Building Secure Wireless LANs**	**141**
Chapter 7	**Planning Wireless LANs**	**143**
	Step 1: Understanding Your Wireless LAN Needs	144
	Step 2: Planning the Scope of Rollout	147
	Step 3: Performing Site Survey	147
	Considering the Geographic Coverage Area	147
	Per-Site Security Requirements	148
	Profiling Wireless LAN Users and Devices	148
	Step 4: Setting Up Requirements and Expectations	149
	Network Bandwidth and Speed	150
	Coverage Area and Range of Wireless LANs	150
	Security	150
	Step 5: Estimating the Required Wireless LAN Hardware and Software	150
	Basic Wireless LAN Hardware	151
	Software	154
	Conventional Hardware Requirements for Various Deployment Scenarios	155
	Step 6: Evaluating the Feasibility of Wireless LANs and the Return on Investment (ROI)	157
	Step 7: Communicating the Final Plan with Higher Executives and Potential Users	158
	An Example of Wireless LAN Planning: Bonanza Corporation	158
	Step 1: Bonanza Wireless LAN Needs	159
	Step 2: Planning the Rollout	160
	Step 3: Site Survey	161
	Step 4: Setting Up Requirements and Expectations	162
	Step 5: Estimating the Required LAN Hardware and Software	163
	Step 6: Evaluating the Feasibility of Wireless LANs and Estimating Return on Investment (ROI)	164
	Step 7: Communicating the Wireless LAN Deployment Plan with Executives	165
	Summary	165
Chapter 8	**Shopping for the Right Equipment**	**167**
	Making Your Wireless LAN Equipment Shopping List	168
	Explore the LAN Technologies Available in the Market	169
	Wireless LAN Technologies	169
	Wired LAN Ethernet Equipment Technologies	169
	Virtual Private Network (VPN) Gateways and Clients	170
	Remote Authentication Dial-in User Service (RADIUS) Server	170
	Wireless LAN Supporting Operating Systems	171
	Major 802.11 Equipment Vendors and their Products	172
	Cisco Systems	172
	Agere Systems/ORiNOCO	174
	Linksys	176

Contents

NetGear	178
Xircom/Intel Corporation	180
Decide Your Shopping Parameters	183
Shopping for LAN Equipment	184
Shopping on the Internet	184
Shopping Using Mail-Order Catalogs	185
Shopping at a Local Computer Hardware or Office Supply Store	186
Shopping Tips	186
Summary	187

Chapter 9 Equipment Provisioning and LAN Setup — 189

Before We Start	190
Identifying the Wireless LAN Components	190
Wireless LAN Adapters	191
Wireless LAN Access Points (APs)	193
Wireless LAN Antennas	193
Networking Support Servers	194
Setting Up a Wireless LAN for the 802.11 Infrastructure Mode	195
Setting Up a Wireless LAN Access Point	195
Setting Up Wireless LAN Adapters	202
Finishing the Access Point Configuration	210
Testing Your Standalone Wireless LAN	215
Adding More Computers to Your Standalone Wireless LAN	216
Connecting a Wireless LAN to the Internet	216
Using Multiple AP Configurations	218
Overlapping AP Configuration	218
Non-Overlapping AP Configuration	220
Setting Up Wireless LAN for the 802.11 Ad-Hoc Mode	222
Summary	223

Chapter 10 Advanced 802.11 Wireless LANs — 225

High Security and Authentication– Enabled 802.11 Wireless LANs	225
The 802.1X Standard	226
Virtual Private Network for Wireless LANs	227
Building a Secure Wireless LAN with 802.1X and VPN Technology	231
Point-to-Point Wireless Connectivity between Two Sites	244
Point-to-Point Wireless Connectivity Requirements.	245
Network Configuration	245
Setting Up ORiNOCO Point-to-Point Radio Backbone Kit	246
Securing the Point-to-Point Wireless Connectivity Using VPN	249
Secure Remote Access from a Wireless LAN over the Internet Using VPNs	249
Summary	250

Contents

PART 4   **Troubleshooting and Keeping Your Wireless LAN Secure**   **251**

Chapter 11   **Troubleshooting Wireless LANs**   **253**
- Common Problems   253
  - Hardware Problems   254
  - Software Problems   256
- Handling Bandwidth Congestion Due to Competing Devices   258
- Upgrading Wireless LANs   259
- Optimizing and Managing the Network Load through Monitoring Wireless LAN Quality   260
- Summary   260

Chapter 12   **Keeping Your Wireless LAN Secure**   **261**
- Establishing Security Policy   262
  - Understanding Your Security Policy Requirements   262
  - Creating Security Policy   265
  - Communicating Security Policy   271
- Security Policy Compliance   271
- Intrusion Detection and Containment   272
  - Wireless LAN AP Monitoring Software   272
  - Intrusion Detection Software   272
  - Antivirus Software   272
  - Firewall and Router Logs   273
  - Network Login and Activity Logs   273
- Getting Ready for Future Security Challenges   273
- Summary   273

Appendix A   **Wireless LAN Case Studies**   **275**
- Home-Based Wireless LANs: The Khwaja Family Residence   276
  - Background   276
  - The Problem   276
  - The Solution   277
  - Results   278
  - Future   278
- A Small Corporation Wireless LAN: The Morristown Financial Group   278
  - Background   278
  - The Problem   279
  - The Solution   279

The Results	279
The Future	280
Campus-Wide Wireless LAN: Carnegie Mellon University	280
Background	280
The Problem	281
The Solution	281
The Results	283
Wireless Internet Service Providers: M-33 Access	283
Background	283
The Problem	283
The Solution	284
The Result	286
The Future	286

Appendix B Installing ORiNOCO PC Card Under Various Operating Systems — 287

Installing under Windows 98, Windows ME, and Windows 2000	287
System Requirements	288
Software Requirements	288
Installation Steps	288
Installing under Windows NT 4.0	294
System Requirements	294
Software Requirements	294
Installation Steps	295
Installing under Mac OS	296
System Requirements	297
Software Requirements	297
Installation Steps	297
Installing under Linux	300
System Requirements	300
Software Requirements	300
Installation Steps	300

Glossary of Terms and Abbreviations — 305

References — 321

Index — 323

About the Authors

Jahanzeb Khan is Principal Engineer with RSA Security, Inc. (formerly RSA Data Security Inc.). He is currently involved in the research and development of Wireless LAN Security standards. At RSA, he is responsible for the research and development of secure network and data communication. Before RSA, he worked at Oracle Corporation and Symantec Corporation, where he was responsible for application software development that required user authentication and security services. Jahanzeb Khan has a B.S. in Computer Science, with emphasis in computer networks and security. He is a member of IEEE International and is active in the 802.11b community. He has over 12 years experience in software and hardware development in general software and computer networks. He has authored various Internet drafts and actively participates in World Wide Web Consortium (W3C) and Internet Engineering Task Force (IETF) activities. He also participates in ongoing discussions relating to Wired Equivalent Privacy (WEP) vulnerability that affects Wi-Fi/802.11 High-Rate Wireless LANs.

Anis Khwaja works in the IT department of a leading financial services firm. He is a long-time veteran of the technology industry and has held leadership position at various technology companies. Prior to his current position, Anis worked as the Director of Technology, Circline Inc. At Circline, Anis was responsible for network infrastructure and software development. He has also worked at CertCo Inc., where he was a development manager responsible for the development of a Public Key Infrastructure (PKI)–based Certificate Authority. Anis has over 15 years of experience in the industry. Previously, he was employed at Attachmate Corporation, where he worked on one of the earliest Internet suites offered by Attachmate. At present, Anis is involved in deployment of 802.11b (Wi-Fi) networks.

Introduction

Wireless connectivity of computing devices is rapidly becoming ubiquitous and soon may be the primary, if not the only, method for many portable devices to connect with computer networks. Wireless LANs provide the easiest way to interconnect computers for both enterprise and SoHo (Small Office, Home Office) environments. First available at airport kiosks, public access has spread through airport waiting rooms, hotels, and restaurants into coffee shops, hospitals, libraries, schools, and other locations. Like any fast growing and successful technology, the phenomenal grown of wireless LANs has been fueled by a convergence of intense customer demand to access data for untethered data access, ever shrinking computing devices, and the standardization of equipment around 802.11b wireless fidelity (Wi-Fi) technology. This has resulted in achieving economies of scale, which enabled prices to go down, further fueling the demand. In this book we explore how secure wireless networks can be built using 802.11 with primary focus on secure wireless LANs.

This book is an implementer's guide to 802.11 (Wi-Fi) wireless networking for home, small offices, enterprises, and Wireless Internet Service Providers (WISPs). It includes introduction and overview of 802.11b (Wi-Fi) technology, planning and design guidelines for implementing wireless LANs, and criteria for evaluating hardware and software. We explore security features and weaknesses, as well as policy management and associated trade-offs in implementing such networks. Quality of service, bandwidth issues, compatibility with related technologies like HomeRF as well as emerging technologies and developments in wireless networking are also examined.

Building Secure Wireless Networks with 802.11 focuses on the wireless LANs that are built using the Institute of Electrical and Electronics Engineers (IEEE) 802.11 standard. The book is a stepwise guide to building a wireless LAN.

First we discuss the basics of wired LANs to help those readers who are either not familiar with LAN technologies and those who would like to gain a better understanding of LANs in general. We talk about the basics of wireless LAN by discussing the primary characteristics of a wireless LAN. We introduce the IEEE 802.11 standards and help you understand the basic differences between the IEEE wireless LAN standards. We also help you evaluate whether wireless LANs are right for you.

One of the primary motivations for writing this book was the fact that the books available at the writing of *Building Secure Wireless Networks with 802.11* did not cover the important security needs of wireless LANs. The authors of this book, given their unique perspective and experience in the computer security industry, recognize security of the wireless LAN as the key factor in determining the future of wireless LANs. In addition to the chapters dedicated to network security, we pay special attention to the security issues of both the wired LANs and that of wireless LANs throughout the book. We discuss standard IEEE 802.11 security as well as the complementary technologies that can be used to provide a robust security to a wireless LAN.

At the end of the book, we also present some real-life case studies to help you visualize the problems that you can solve using a wireless LAN, the challenges that you might face, and the outcomes of using a wireless LAN.

Who Should Read This Book

The book in its entirety best serves individuals and information architects who want to create and use wireless LAN solutions. The readers of the book could be home users who want to connect multiple computers at home using the wireless LANs; SoHo network administrators or users who want the mobility provided by the wireless LANs; and the Enterprise IT managers and architects who want to deploy secure wireless LANs and need to understand the issues surrounding wireless LANs. *Building Secure Wireless Networks with 802.11* is where you can find the plain-English information you need to put Wireless LANs to work.

What You Need to Know

Every book ever written makes some basic assumptions about the reader; some require a user to have in-depth knowledge of the subject, whereas others could be written with a layman in mind. *Building Secure Wireless Networks with 802.11* is written for readers who may have different levels of knowledge and understanding of wireless LANs. The book starts from the very basics of LAN technologies and extends the discussion to the latest available wireless LAN technologies. The book attempts to build a foundation that can help you feel comfortable exploring more information on subjects that might not be covered in this book.

We do, however, recommend that you have some basic knowledge of networking concepts, TCP/IP, as well as familiarity with the software networking components of the Microsoft Windows operating systems. Any such knowledge will help you grasp the ideas discussed in this book at a faster pace.

How This Book Is Organized

Building Secure Wireless Networks with 802.11 contains a wealth of information that you can put to work right away. This book presents a step-by-step approach for understanding and implementing a Wireless LAN based on 802.11b (Wi-Fi) technology. It includes detailed information on every aspect of setting up, configuring, and managing your wireless LAN. The book is divided into four parts for better organization and readability.

Part 1, "Introduction to Wireless Local Area Networks (LANs)," first explains basic networking, wireless networking, and IEEE 802.11 wireless standards, and then provides you with the baseline, which will allow you to decide whether wireless LANs are right for you. It has four chapters.

- Chapter 1, "Networking Basics," talks about the history of computer networks and describes different types of computer networks, as well as different topologies and networking hardware and the principles behind them. We briefly discuss the International Standards Organization Open Systems Interconnection (ISO/OSI) Reference Model and its significance in the development of network standards.

- Chapter 2, "Wireless LANs," explains the basic design and operation of wireless LANs. We explore the basics of wireless networks and look into a brief history of wireless networks. We first outline the basics of wireless networks, then we study the wireless LAN architecture in detail and the technologies that constitute a wireless LAN.

- In Chapter 3, "The Institute of Electrical and Electronics Engineers (IEEE) 802.11 Standards," we examine both the approved and up-and-coming wireless LAN standards of the Institute of Electrical and Electronics Engineers (IEEE). Our focus will be the 802.11 standard proposed by the wireless LAN working group. We will explain the differences between various 802.11 standards, their operation, interoperability, and deployment constraints.

- Chapter 4, "Is Wireless LAN Right for You?" helps you decide whether a wireless LAN is right for you. We discuss the different aspects of a wireless LAN that directly impact the deployment feasibility in SoHo, Enterprise, and Wireless Internet Service Provider scenarios. We talk about the benefits, deployment scenarios, costs associated, deployment issues, bandwidth and network congestion, security, and health concerns of the wireless LANs.

Part 2, "Secure Wireless LANs," first discusses the security issues of wired LANs, then continues to talk about the security issues of wireless LANs and how to secure them. It has two chapters.

- Chapter 5, "Network Security," clarifies the basics of network security by discussing the different types of network security, commonly known attacks against computer networks, and the most common practices that are used to ensure security of a LAN.

- Chapter 6, "Securing the IEEE 802.11 Wireless LANs," examines the special security requirements of a wireless LAN. It provides a brief overview of security primitives in the IEEE 802.11 standard. We explore the weaknesses in the current security model that 802.11 standard compliant devices use. We also discuss the additional security measures that can be used in 802.11 standard based LANs to provide a higher level of security than defined in the standard.

Part 3, "Building Secure Wireless LANs," helps you build a real-world wireless LAN. First we help you plan a wireless LAN, then we help you choose the right equipment for your deployment scenario. We also guide you through the steps with the equipment provisioning. Finally, we discuss how to connect a wireless LAN with a remote network using VPNs. Part 3 has four chapters.

- Chapter 7, "Planning Wireless LANs," explains the significance of planning a wireless LAN. We help you make the basic decisions that help you build an extensible and flexible wireless LAN.

- Chapter 8, "Shopping for the Right Equipment," helps you decide what kind of wireless LAN equipment you will need for a particular deployment scenario. We talk about equipment selection based on SoHo, Enterprise, and WISP scenarios.

- Chapter 9, "Equipment Provisioning and LAN Setup," discusses the actual process of setting up wireless LANs. In this chapter we help you design a wireless LAN that provides a secure operation and suits your needs.
- Chapter 10, "Advanced 802.11 Wireless LANs," explains how to extend a wireless LAN by connecting it with an enterprise LAN using a virtual private network (VPN) and the 802.1x authentication protocol.

Part 4, "Troubleshooting and Keeping Your Wireless LAN Secure," details the issues in maintaining and troubleshooting a wireless LAN. Part 4 has two chapters.

- Chapter 11, "Troubleshooting Wireless LANs," discusses some of the common issues surrounding the troubleshooting and maintenance of a wireless LAN. These issues include the common problems, handling bandwidth congestion due to competing devices, upgrading wireless LAN equipment, and optimizing and managing network overload through monitoring.
- Chapter 12, "Keeping Your Wireless LAN Secure," talks about developing practical wireless LAN security policies that work. We discuss the process of developing and establishing wireless LAN security policies and how to integrate them into an organization.

It is the sincere hope of the authors that this book will help you understand the wireless LAN technology in general, the IEEE 802.11 standards, the wireless LAN security requirements and solutions to the current security weaknesses to successfully build a secure wireless LAN. As the awareness of wireless LAN technologies grows, so will the importance and significance of wireless LANs and its tools, which will in turn be reflected in the future wireless LANs. Perhaps with the right combination of awareness, newer and better technologies, and cost effectiveness, wireless LANs will soon become ubiquitous, redefining the way we use computers today.

PART One

Introduction to Wireless Local Area Networks (LANs)

Wireless local area networks (LANs) are a new breed of LANs that use airwaves instead of a physical medium (wires or cables) to interconnect computers. Though wireless LANs use many of the same fundamental principles that wired LANs do, wireless LANs need a lot more attention when it comes to their deployment. In order to successfully deploy wireless LANs, you must understand the basics of a wired LAN and that of the wireless LANs. You must carefully choose a standard-based wireless LAN technology that would be upwardly compatible with future standards. You should consider the pros and cons of wireless LANs before you deploy them to ensure that wireless LANs are right for you. Part 1 of this book talks about all these issues by walking you through the basics of wired and wireless networks, the prevalent standards, and pros and cons of wireless LANs.

Chapter 1 talks about the history of computer networks, describes different types of computer networks, and discusses the different topologies and networking hardware and the principles behind them. We briefly discuss the International Standards Organization Open Systems Interconnection (ISO/OSI) Reference Model and its significance in network equipment standards development.

Chapter 2 explains the basic design and operation of wireless LANs. We explore the basics of wireless networks and talk about a brief history of wireless networks. We go over what a basic wireless network consists of, then we study wireless LAN architecture in detail and the technologies that make up a wireless LAN.

In Chapter 3, we examine the wireless standards that Institute of Electrical and Electronics Engineers (IEEE) 802 Local Area Network and Metropolitan Area Network Standards Committee (LMSC) committee has approved and those that are up and coming. Our focus will be 802.11, the wireless LAN working group. We will understand the differences between various 802.11 standards, their operation, interoperability, and deployment constraints.

Wireless LANs are relatively new technology. They have some great benefits and few known weaknesses. Chapter 4 helps you decide whether wireless LAN is right for you. We discuss the different aspects of a wireless LAN that directly impact the feasibility for Small Office Home Office (SoHo), Enterprise, and Wireless Internet Service Provider (WISP) deployment scenarios. We talk about the benefits, deployment scenarios, costs associated, deployment issues, bandwidth and network congestion, security, and health concerns of the wireless LANs.

It is likely that you are already familiar with the basic terminology, devices, and principles associated with LANs—history of wired and wireless LANs, network interface cards, wireless network operation, and so on—equally, there is a fundamental set of techniques and terminology associated with wireless LANs and these are often less well understood. When you finish reading Part 1, you will understand the evolution of wireless LANs and LANs in general. You will be able to understand basic wireless LAN operation and the industry standards that wireless LANs are following today. You will be able to identify the pros and cons of using wireless LANs and assess whether wireless LAN is right for you.

CHAPTER 1

Networking Basics

Over the last ten years computer networks have increasingly become part of our daily lives. From the Internet (which is a network of networks) to networks at work, grocery stores, video stores, banks, and hospitals, almost every place seems to be connected with some sort of computer network. A basic computer network is formed when two or more computers are connected together to share processing power and resources or to intercommunicate for other reasons. For example, a computer network at work interconnects various computers to facilitate cooperation among employees through file sharing, email messaging, application programs, and data management. At stores, computers work together to provide detailed information about product availability, pricing, and shipment. Banks use computer networks to perform account management functions where accurate data management is extremely important. Just imagine if all these places had only one computer performing all these tasks! We all might have to wait in lines for hours before we got served.

The computers that are only interconnected at a given premises are said to be operating in a local area network (LAN) environment. Often these networks are connected with other networks or the Internet to provide instant access to more information. However, sometimes for security reasons, LANs are restricted to local and private access only.

In this chapter, we go over the history of computer networks, describe different types of computer networks, talk about the different topologies and networking hardware and the principles behind them, and we introduce the Transmission Control Protocol/Internet Protocol (TCP/IP) network protocol and its basic parameters. At the end of this chapter, we put together an example that walks you through the process of setting up a hypothetical LAN.

Development of Computer Networks: An Overview

On September 11, 1940, George Steblitz used a Teletype machine at Dartmouth College in New Hampshire to transmit a problem to his Complex Number Calculator in New York and received the results of the calculation on his Teletype terminal. This round-trip transfer of data is considered the first example of a computer network. Later, in 1958, the second computer network was unveiled at the Massachusetts Institute of Technology (MIT) based on the time-sharing technology called Project MAC (for Multiple Access Computer and Machine-Aided Cognition). Time-sharing technology is basically the rapid time-division multiplexing of a central processor unit (CPU) among the jobs of several users, each of which is connected with the CPU using a typewriter-like console. Time-sharing computer systems allow multiple simultaneous users the ability to share the CPU time among them while giving to each of them the illusion of having the whole machine at his or her disposal. Project MAC developed the Compatible Time-Sharing System (CTSS), one of the first time-shared systems in the world, and Multics, an improved time-shared system that introduced several new concepts. These two major developments stimulated research activities in the application of online computing to such diverse disciplines as engineering, architecture, mathematics, biology, medicine, library science, and management. CTSS was first demonstrated in 1961, and it included facilities for editing, compiling, debugging, and running in one continuous interactive session that has had the greatest effect on programming. Prior to CTSS, computer systems had extremely cumbersome programming environments. For example, a programmer had to load an entire program into a CPU using a punch card or keyboard every time he or she wanted to test or make minor changes to the program. The availability of programming facilities in project MAC enabled professional programmers to be more imaginative in their work and to investigate new programming techniques and new problem approaches because of the much smaller penalty for failure. International Business Machines (IBM) and General Electric (GE) were the major sponsors of project MAC.

On April 7, 1964, IBM introduced the System/360 that included a Time Share System (TSS) based on CTSS. In 1969, Bell Labs announced its own network-aware computer operating system called UNIX. UNIX included built-in support for networking computers. UNIX offered a practical solution to interconnecting computer systems to form local area networks.

Realizing the growing need for interconnecting separate computer networks, that same year the Department of Defense (DOD) launched its private network called ARPANET. ARPANET, now known as the Internet, was brought online in December 1969 as a wide area network (WAN) that initially connected four major computers at universities in the southwestern United States (UCLA, Stanford Research Institute, UCSB, and the University of Utah), and it was strictly restricted for research use. ARPANET became extremely popular among researchers in both government and the scientific community, and many other research facilities and universities were added to the ARPANET.

By the late 1960s, advancement in computer systems reduced the size of the computers and enhanced the computing power. The computers that took up a room in the early 1960s could now fit into a space the size of a large filing cabinet. These newer and smaller computers were called minicomputers. These computers were rapidly adopted by commercial organizations, and computers were deployed not only for complex computations but to provide business solutions to organizations. With greater computation needs, having more than one computer on the premises in large organizations was not unrealistic. Such computers were connected to one another to share resources like printers and punch-card readers and perform complicated tasks using application programs. These application programs performed tasks ranging from complicated mathematical calculations to keeping bank records. This distributed computation environment where multiple computers and peripherals needed to communicate with each other required a data communications network to tie the computer systems with the peripherals to form LANs. These LANs needed to have high bandwidth. In fact, LANs had to accommodate speeds that were orders of magnitude greater than the original time-sharing networks. Entire application programs had to be downloaded to multiple users. Files, the results of running applications program, had to be uploaded to be stored in central memory.

Robert Metcalfe was a member of the research staff for Xerox at their Palo Alto Research Center (PARC), where some of the first personal computers were being made. Metcalfe was asked to build a networking system for PARC's computers. Xerox's motivation for the computer network was that they were also building the world's first laser printer and wanted all of PARC's computers to be able to print using this printer. The news media have often stated that Ethernet, the most widely used network protocol, was invented on May 22, 1973, when Metcalfe wrote a memo to his bosses stating the possibilities of Ethernet's potential, but Metcalfe claims Ethernet was actually invented very gradually over a period of several years. In 1976, Robert Metcalfe and his assistant, David Boggs, published a paper titled "Ethernet: Distributed Packet-Switching for Local Computer Networks." The object of Ethernet was to design a communication system that was inexpensive and could grow smoothly to accommodate several buildings full of computers. The paper talked about an experience of using 100 computers with a combined wiring extending up to 1 kilometer long coaxial cable. Consequently, Metcalfe and Boggs chose to distribute control of the communications facility among the communicating computers to eliminate the reliability problems of an active central controller, to avoid creating a bottleneck in a system rich in parallelism so that the failure of a computer tended to affect the communications of a computer instead of making the entire network unusable, and to reduce the fixed costs that make small systems uneconomical. The most important innovation of this paper was the absence of a central control—"An Ethernet's shared communication facility, its Ether, is a passive broadcast medium with no central control" (Metcalfe)—which had been the most commonly used method of controlling network traffic before Ethernet. This choice, to make Ethernet relatively inexpensive to build, maintain, and deploy, has been a key factor in its later adoption and success. IBM initially defined the Token Ring at its research facility in Zurich, Switzerland, in the early 1980s. Computers on a Token Ring LAN are organized in a ring topology (see the section titled *Ring Topology* later in this chapter) with data being transmitted sequentially from one ring station to the next. IBM pursued standardization of Token Ring under the 802.5 Working Group of the Institute of Electrical and Electronics Engineers (IEEE). Today, Token Ring is the second most widely used LAN technology. Token Ring LANs provided higher speed than Ethernet, but they are far more costly than Ethernet. Personal computers (PCs) were the revolution of the mid-1970s. Many consider Altair 8800 released by Micro Instrumentation and Telemetry Systems, Inc. (MITS) in 1975 to be the first PC. In 1977, Apple Computers, Inc. introduced the Apple II, a PC with a color monitor, sound, and graphics. In 1977, Dennis Hayes invented a device called modulator demodulator (MODEM), which enabled computers to communicate with one another over the regular phone line. In 1980, IBM introduced the IBM PC, which soon became a standard in the enterprise market. PCs were much smaller in size than their predecessor minicomputers and the mainframes. PCs were small enough to be placed on a desk, whereas minicomputers still required at least an area equivalent to a refrigerator. In addition to their size, PCs were much cheaper and faster than their rival minicomputers. Companies rapidly started replacing old and noisy typewriters with quieter and slicker PCs. The networking equipment and standards were already present when PCs arrived in the market. LANs started proliferating within organizations.

During the 1980s, while the speed of LANs and PCs kept on growing, there was an increased interest among organizations in communicating with other organizations and interconnecting their offices using computers; meanwhile computer enthusiasts were also interested in reaching out to other computer users. Organizations and individuals started setting up bulletin board systems (BBS), which used modems and phone lines to connect to other computers, to communicate with their customers and individuals. BBSs offered a low-cost solution for sharing files. BBS systems provided a computer terminal look and feel to remote computers. A BBS system consists of a PC equipped with one or more modems each connected with a phone line using BBS communication software. A user willing to access the BBS needed a PC, a modem, and a phone line with appropriate BBS software. BBS systems were not very secure, however, and were extremely vulnerable to malicious attacks from hackers who tried to degrade the performance of BBS systems by keeping the system busy, and to fill up the disk space on BBS systems by uploading unnecessary files.

The growing need for a public data network was becoming clear, and in 1983 ARPANET was split into ARPANET and MILNET; the latter became integrated with the Defense Data Network (DOD private network). In 1986, the National Science Foundation funded NSFNet as a cross-country 56 Kbps backbone for the Internet. November 3, 1988, is known by many computer enthusiasts as Black Thursday. On this day, a computer virus, known as the worm, burrowed through the Internet, affecting almost 6,000 of the 60,000 hosts on the Internet. The growing demand for the NFSNet and ARPANET kept on increasing, and ARPANET finally decommissioned in 1989. NSF gave control of NFSNet to the private sector, allowing commercial use of NFSNet, the remaining ARPANET, and any commercial extensions of the Internet. The development of the Internet took off once it was allowed to be used commercially. In 1991, the World Wide Web (WWW) was released by the European Organization for Nuclear Research (CERN), changing the way we live our lives today.

The advancements in silicon-chip technology facilitated increased network speed. Computer networks started operating at higher and higher speeds. The physical medium was improved, the protocols were enhanced, and smaller network devices were designed that consumed less power and were more reliable. Today, most LANs use the Ethernet adapters and operate at speeds in the range of 10 to 100 megabits per second (Mbps). These LANs are normally connected to other bigger networks or Internets via broadband connections or private lines using asynchronous transfer mode (ATM), Frame Relay, or other technologies. ATM and Frame Relay are high-performance WAN protocols that share a transmission medium and are normally used in situations where a reliable network connectivity is desired.

Even with these advancements in computer networking, there is room for higher network speeds. Standards organizations and research labs are constantly working on developing even faster computers and the networks to connect them.

Network Types

Computers can be networked in many different ways, forming different types of networks. The networking type is normally determined by the intended use, size, and geography of the computers on the network. Some of the examples of different network types are peer-to-peer networks, local area networks, wide area networks, personal area networks, virtual private networks, and the Internet.

Peer-to-Peer Networks

A peer-to-peer network consists of two or more computers that are directly connected to one another (see Figure 1.1). Such computer networks are normally insecure and operate at higher speeds than other types of networks. However, peer-to-peer computer networks are usually not very flexible and have limited scope. Peer-to-peer networks are considered to be operating in secure environment if the peers (computers in the network) mutually trust each other and there is no fear of a successful intrusion by an adversary.

An example of a peer-to-peer network might be a home computer network or a home office computer network, where two or more computers are interconnected to share files or computer processing power.

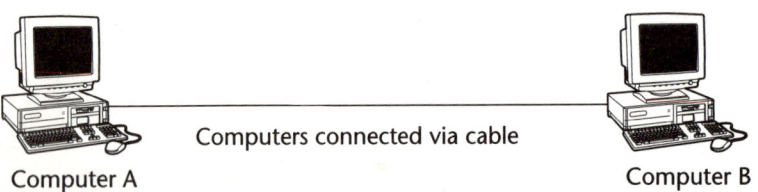

Figure 1.1 Peer-to-Peer Network

Local Area Networks (LANs)

Local area networks enable computers to share processing power, files, and other resources like printing services. LANs are normally deployed in places where certain LAN services (file sharing or printing) are required to be reliable (see Figure 1.2). In most cases, LANs contain one or more file servers (computers with large hard drives for sharing files), print servers (for sharing printers), and authentication servers (to ensure that only authorized people can use the shared services). All the computers sharing the resources on a network must be configured with the protocols used by the LAN. Most LANs today use TCP/IP as the higher-level protocol; with Ethernet adapters that are physically connected to the network using twisted pair cabling. Most private LANs (a network that is not accessible by the outside world) are secured, but they are still vulnerable to a host of influences, from honest mistakes by employees running a software virus on their computers to disgruntled employees who intentionally target a company's information assets.

Wide Area Networks (WANs)

Depending on the technology used, LANs normally have a geographic limit of 100 meters. This is restrictive in terms of connecting two offices, which might be in two different cities. Wide area networks (WANs) take connectivity to a much higher level by enabling computers to connect with other computers or networks at much farther distances. A computer may be connected to a LAN thousands of miles away in a different city or perhaps a different continent. Two different LANs might be interconnected using a WAN link, which can exist over a phone line or a private leased line (see Figure 1.3). A WAN link is like a road between one place and another, busy place. The data exchanged over a WAN link is not considered to be secure unless it is transferred in an encrypted format (that is, data is encrypted before it is sent, and it is decrypted by the intended recipient upon receipt).

Today, WAN links are widely used and enable companies and individuals to stay connected and provide location transparency.

10 **Chapter 1**

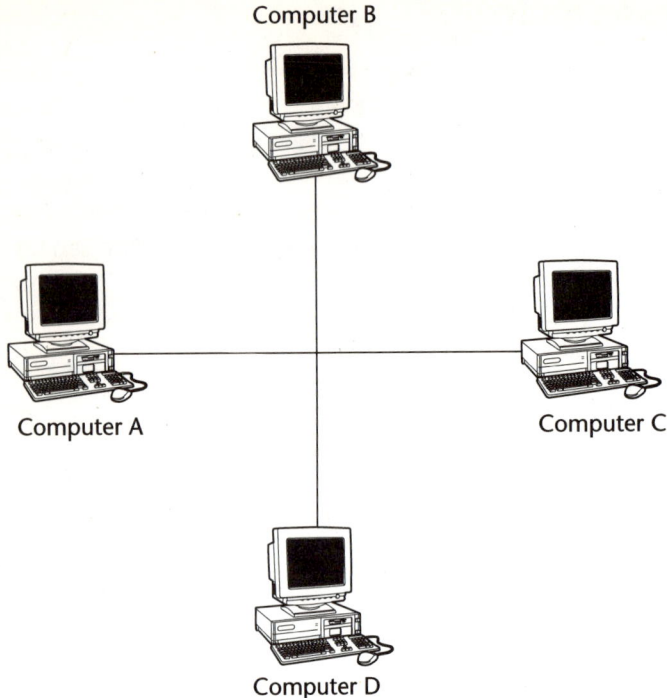

Figure 1.2 LAN with more than two computers.

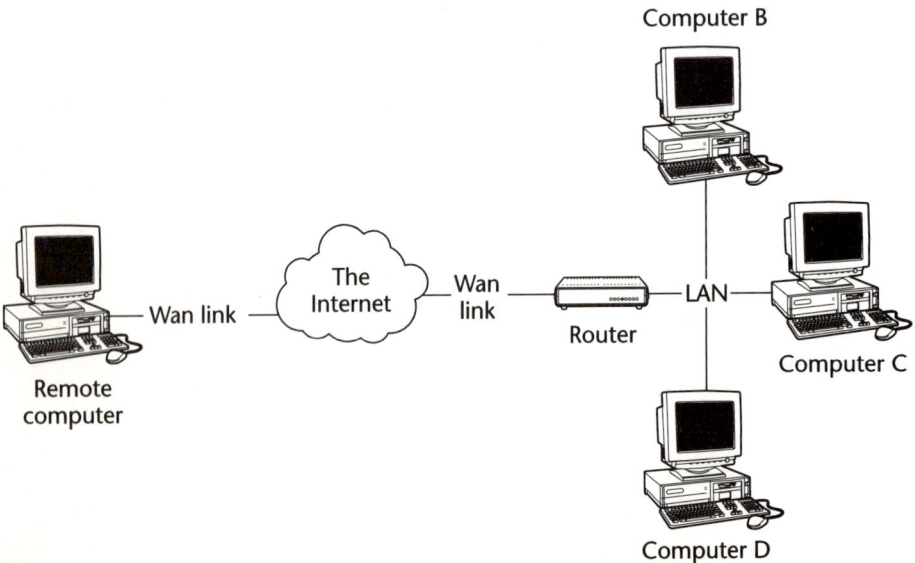

Figure 1.3 WAN link.

Personal Area Networks (PANs)

Personal area networks (PANs) are extremely low power, normally wireless, communication devices that enable a PAN-enabled device to exchange data with a PAN-aware device within a short distance (see Figure 1.4). Examples of such devices include handheld personal digital assistants (PDAs), human authentication devices, and payment systems. PANs are relatively new to the market. Lots of work is being done in this area to provide a higher level of information sharing and personal security.

The Internet

The Internet in all its guises, permutations, and uses is extremely complex. But basically the Internet can be defined as a network of computer networks (see Figure 1.5). It can be thought of as a tree, where the Internet itself is the main trunk, networks connected to the Internet are branches, and the leaves on the branches are the computers on the Internet. The Internet uses TCP/IP as the protocol for exchanging data and information. In physical terms, the Internet is a global mesh of high-performance, high-bandwidth communications infrastructure consisting of a variety of communication equipment and connecting links (for example, copper cable, optical cables, satellites, and so on) together known as the Internet backbone. Access to this high-speed backbone is controlled by the major communication providers, which provide the access to the Internet Service Providers (ISPs). These ISPs resell the access to individuals and corporations for connectivity. This enables anyone with access to the Internet to reach anyone else who is also connected to the Internet.

The level of connectivity provided by the Internet has boosted the economy worldwide. Internet merchandising, emails, news, personal communication, and remote connectivity have changed the way we live today.

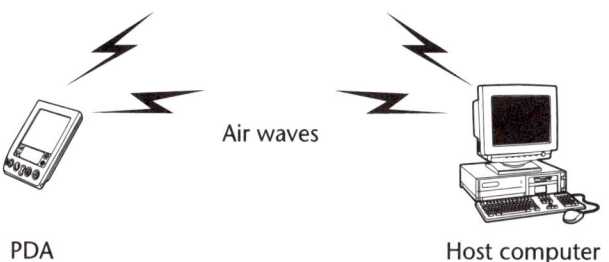

Figure 1.4 PDA used in conjunction with a PC.

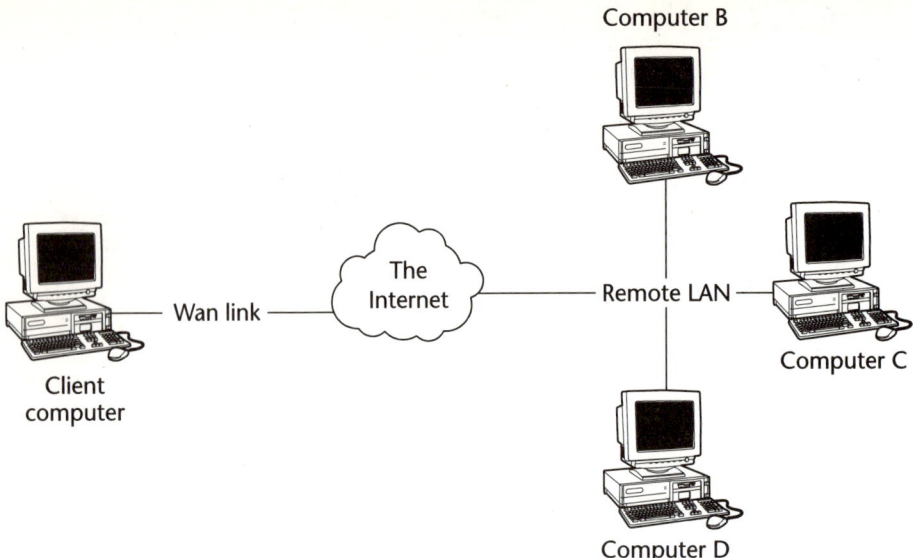

Figure 1.5 Simple rendering of Internet showing a desktop computer accessing a remote network.

Virtual Private Networks (VPNs)

Virtual private networks (VPNs; see Figure 1.6) are an extension of WANs. As mentioned earlier, WANs allow a computer to be connected to a remote LAN via a WAN link (where a WAN link can be over a phone line or a private leased line). The data exchanged over a WAN link can go through many computers and provide hackers and adversaries with a chance to eavesdrop and access this information, even altering it or using it for profit. A secure tunnel between the computer and the remote LAN is required to protect the information. The VPNs fit this requirement by allowing only authorized personnel access to the LAN. All the data is exchanged in an encrypted format so that it cannot be eavesdropped upon.

VPNs are becoming extremely popular. Most organizations that allow their employees to work remotely use a VPN connection over a WAN link instead of a raw WAN connection.

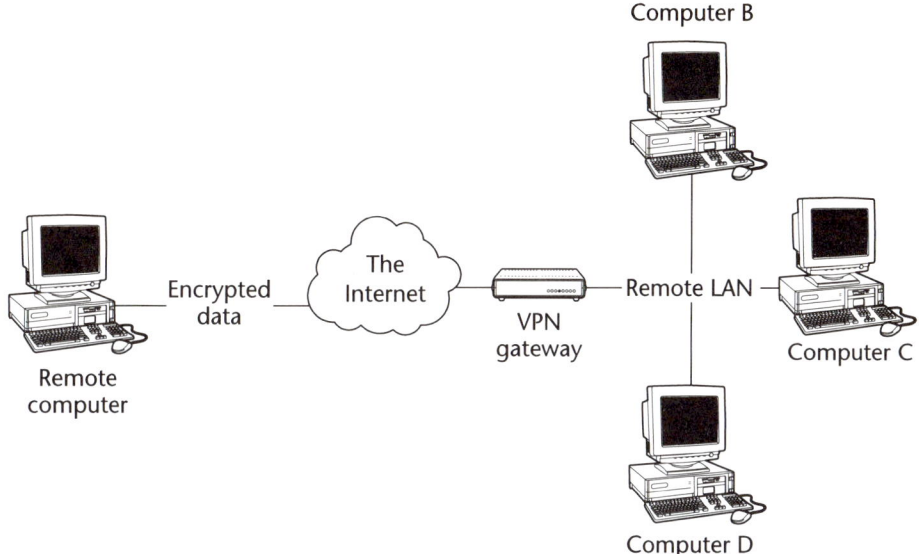

Figure 1.6 VPN connected to the Internet.

Network Topologies

Network topology refers to the shape of a network, or the network's layout. How different computers in a network are connected to each other and how they communicate is determined by the network's topology.

Three Commonly Used Topologies

The computers on a network can be arranged in many different ways, but the most commonly used topologies are bus, ring, and star.

Bus Topology

In a bus topology, all the devices are connected to a central cable (see Figure 1.7). It is the most commonly used network topology, having various adaptations, among them linear bus, bus with extensive branching, and bus tree. These adaptations came about with specified electrical properties that allow longer drops and drops within drops. With all bus topologies, communications are conducted on common conductors where the receiver and transmitter are connected to the same communication wires as all other network nodes. This allows the transmission from one node to be received by all others.

In a bus topology all the devices have simultaneous access to the bus. The computer network must use a protocol to control such access to avoid collision and corruption of data. The most common type of such a protocol is Carrier Sense Multiple Access with Collision Detection (CSMA/CD), or Ethernet.

Ring Topology

The second most popular network topology is ring topology, in which each node acts as a repeater (see Figure 1.8). Transmission starts at a central station, usually the controller, and is sent to one node. That node receives the transmission, processes the information if needed, and then sends it to the next node on the ring. Long networks are possible because each node reconditions the transmission, and throughput time around the ring is predictable. When the ring breaks, communication is lost; hot swapping is not possible (a new node cannot be inserted in the ring while the network is in operation). All devices are connected to one another in the shape of a closed loop, so that each device is connected directly to two other devices, one on either side of it

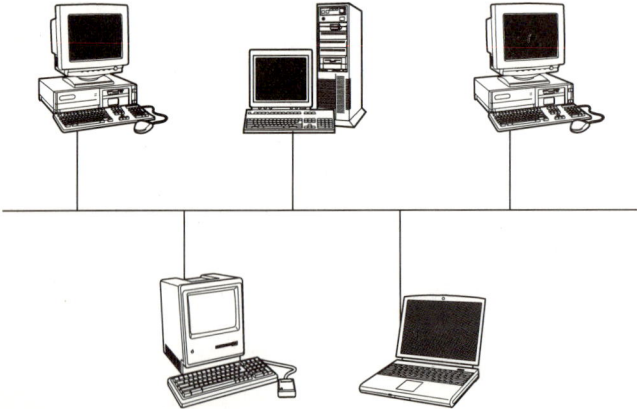

Figure 1.7 Bus topology.

Networking Basics 15

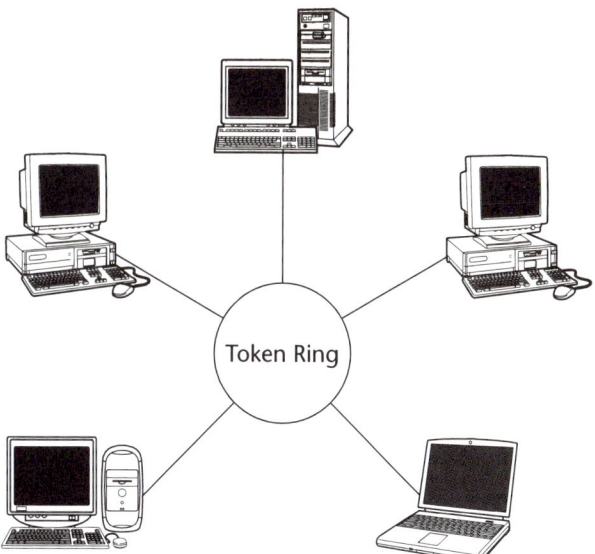

Figure 1.8 Ring topology.

Ring topology provides a high throughput and is normally used to construct corporate LAN backbones.

Star Topology

In a star topology, all devices are connected to a central hub (see Figure 1.9). Nodes communicate across the network by passing data through the hub. Because the protocol is easy to develop, many private networks use it. The mesh topology connects each node with every other node, creating an isolated data path between each node.

Star topology has a very high performance but works in a limited geographical area and is very costly, as the wires from each computer must run all the way to the central hub. Most wireless networks use a variation of the star topology (without wires, of course).

Choosing the Right Topology

Which topology you deploy should be based upon connectivity requirements, budget, and the available hardware. The bus topology is the simplest to implement and is the most widely used network topology. The ring topology is the most expensive to implement. Bus topology is extremely common in enterprise LANs; however, their backbones are often designed using the ring topology to give higher performance. Ring topology attains better performance over bus topology because the physical medium that data travels on is not shared among all computers on the network (only adjacent computers share the given medium), whereas in bus topology all computers connected to the network share the same physical medium, resulting in collision and medium congestion (network becomes too busy) and hence lower performance. Wireless LANs use the star topology because it provides a better management of the network bandwidth.

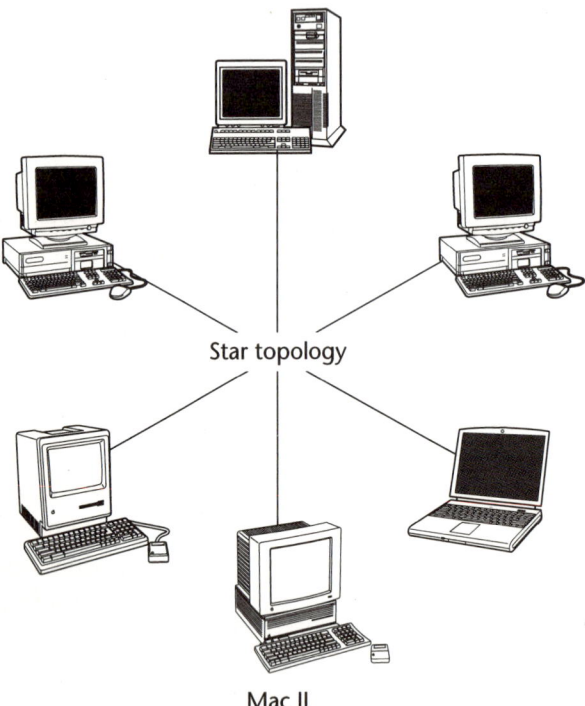

Figure 1.9 Star topology.

Network Hardware and Software

In this section we talk about the networking components, software, and the protocols that are required for each computer in a network. For a network to function, all the computers must have compatible network software and hardware, and they must be connected to one another via a physical link, a cable, for example.

Networking Components

A computer in a network must have a network interface card (NIC) installed. These are electronic circuits that conform to the physical layer of the International Standards Organization Open Systems Interconnection (ISO/OSI) Reference Model and are IEEE-compliant. These network cards connect the computer to a network. In this section we discuss the ISO/OSI Reference Model and the IEEE view of the first two layers of this model. We also discuss NICs, hubs, routers, and repeaters.

International Standards Organization Open Systems Interconnection (ISO/OSI) Reference Model

Modern computer networks are designed in a highly structured way. To reduce the design complexity, most networks are organized as a series of layers, each one built upon its predecessor.

The ISO/OSI Reference Model (Figure 1.10) is based on a proposal developed by the International Standards Organization (ISO). The model is called ISO/OSI Reference Model because it deals with connecting open systems—that is, systems that are open for communication with other systems.

Flexibility is the primary requirement for an acceptable open system. Prior to ISO/OSI Reference Model, most computer networks were proprietary and monolithic (you had to buy the entire network system from one vendor). They were not interoperable with other network systems and were hard to maintain. The ISO/OSI Reference Model added flexibility to the network model by dividing a network system into seven distinct parts. Control is passed from one layer to the next, starting at the application layer, proceeding to the bottom layers. Since the seven layers are stacked on top of one another, the reference model is also known as ISO/OSI stack. The reference model allows different vendors to manufacture networking components that interoperate with each other and hence provides a better option to a network implementer who can build a network based upon his or her need. For example, today we use HyperText Transfer Protocol (HTTP) to surf the Internet. Let's assume that starting next week you would have to use a new protocol called ViperText Transfer Protocol (VTTP). If the protocol is written with ISO/OSI Reference Model in mind, all you would have to do is to install the VTTP protocol driver and you would be ready to use the VTTP without any other modification to your network hardware or software. The principles that were applied to arrive at the seven layers are as follows:

1. A layer should be created where a different level of abstraction is needed.
2. Each layer should perform a well-defined function.
3. The function of each layer should be chosen to interoperate with internationally standardized protocols.
4. The number of layers should be large enough that distinct functions need not be thrown together in the same layer out of necessity, and small enough that the architecture does not become unwieldy.

The computer systems that implement their network components using the ISO/OSI Reference Model can interoperate with most other systems. A layer can be replaced with another layer of the same type from a different vendor. This provides great flexibility to systems manufacturers, IT staff, and general users where they can plug and play different protocols, adapters, and networks without making drastic changes on their computers.

Now let's look at the layers that the OSI Reference Model defines.

The Application Layer: Layer 7

The application layer contains a variety of protocols that are commonly needed. For example, there are hundreds of incompatible terminal types in the world. Consider the plight of a full-screen editor that is supposed to work over a network with many different terminal types, each with different screen layouts, escape sequences for inserting and deleting text, ways of moving the cursor, and so on.

One way to solve this problem is to define an abstract network virtual terminal for which editors and other programs can be written. To handle each terminal type, a piece of software must be written to map the functions of the network virtual terminal onto the real terminal. For example, when the editor moves the virtual terminal's cursor to the upper left-hand corner of the screen, this software must issue the proper command sequence to the real terminal to get its cursor there too. All the virtual terminal software is in the application layer.

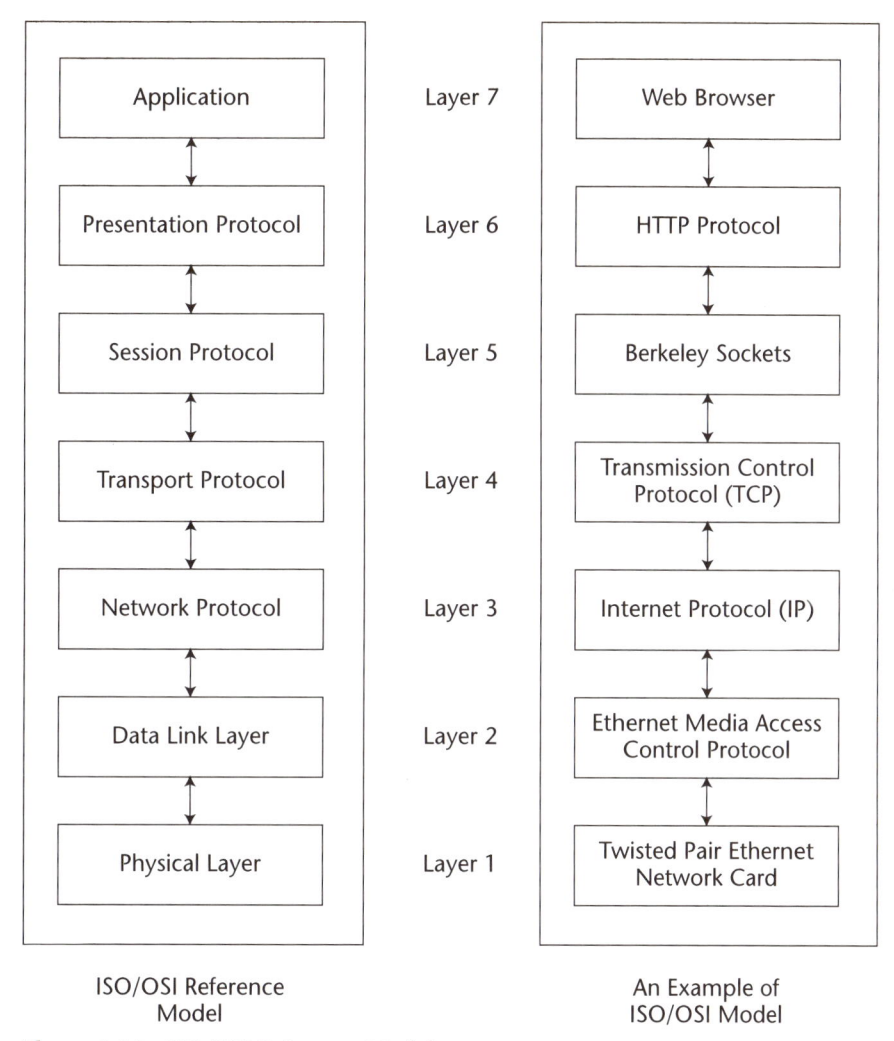

Figure 1.10 ISO/OSI Reference Model.

Another application layer function is file transfer. Different file systems have different file-naming conventions, different ways of representing text lines, and so on. Transferring a file between two different systems requires handling these and other incompatibilities. This work, too, belongs to the application layer, as do electronic mail, remote job entry, directory lookup, and various other general-purpose and special-purpose facilities.

The Presentation Layer: Layer 6

The presentation layer performs certain functions that are requested sufficiently often to warrant finding a general solution for them, rather than letting each user solve the problems. In particular, unlike all the lower layers, which are just interested in moving bits reliably from here to there, the presentation layer is concerned with the syntax and semantics of the information transmitted.

A typical example of a presentation service is encoding data in a standard, agreed-upon way. Most user programs do not exchange random binary bit strings. They exchange things such as people's names, dates, amounts of money, and invoices. These items are represented as character strings, integers, floating-point numbers, and data structures composed of several simpler items. Different computers have different codes for representing character strings, integers, and so on. In order to make it possible for computers with different representations to communicate, the data structures to be exchanged can be defined in an abstract way, along with a standard encoding to be used "on the wire." The job of managing these abstract data structures and converting from the representation used inside the computer to the network standard representation is handled by the presentation layer.

The presentation layer is also concerned with other aspects of information representation. For example, data compression can be used here to reduce the number of bits that have to be transmitted, and cryptography is frequently required for privacy and authentication.

The Session Layer: Layer 5

The session layer allows users on different machines to establish sessions between them. A session allows ordinary data transport, as does the transport layer, but it also provides some enhanced services useful to an application. A session might be used to allow a user to log into a remote time-sharing system or to transfer a file between two machines.

One of the services of the session layer is to manage dialogue control. Sessions can allow traffic to go in both directions at the same time, or in only one direction at a time. If traffic can go only one way at a time, the session layer can help keep track of whose turn it is.

A related session service is token management. For some protocols, it is essential that both sides do not attempt the same operation at the same time. To manage these activities, the session layer provides tokens that can be exchanged. Only the side holding the token may perform the critical operation.

Another session service is synchronization. Consider the problems that might occur when trying to do a two-hour file transfer between two machines on a network with a one-hour mean time between crashes. After each transfer was aborted, the whole transfer would have to start over again, and would probably fail again with the next network crash. To eliminate this problem, the session layer provides a way to insert checkpoints into the data stream, so that after a crash, only the data after the last checkpoint has to be repeated.

The Transport Layer: Layer 4

The basic function of the transport layer is to accept data from the session layer, split it up into smaller units if need be, pass these to the network layer, and ensure that the pieces all arrive correctly at the other end. Furthermore, all this must be done efficiently and in a way that isolates the session layer from the inevitable changes in the hardware technology.

Under normal conditions, the transport layer creates a distinct network connection for each transport connection required by the session layer. If the transport connection requires a high throughput, however, the transport layer might create multiple network connections, dividing the data among the network connections to improve throughput. On the other hand, if creating or maintaining a network connection is expensive, the transport layer might multiplex several transport connections onto the same network connection to reduce the cost. In all cases, the transport layer is required to make the multiplexing transparent to the session layer.

The transport layer also determines what type of service to provide to the session layer, and ultimately, the users of the network. The most popular type of transport connection is an error-free point-to-point channel that delivers messages in the order in which they were sent. However, we have other possible kinds of transport, service, and transport-isolated messages with no guarantee about the order of delivery, and broadcasting of messages to multiple destinations. The type of service is determined when the connection is established.

The transport layer is a true source-to-destination or end-to-end layer. In other words, a program on the source machine carries on a conversation with a similar program on the destination machine, using the message headers and control messages.

Many hosts are multiprogrammed, which implies that multiple connections will be entering and leaving each host. There needs to be a way to tell which message belongs to which connection. The transport header is one place this information could be put.

In addition to multiplexing several message streams onto one channel, the transport layer must take care of establishing and deleting connections across the network. This requires some kind of naming mechanism so that a process on one machine has a way of describing with whom it wishes to converse. There must also be a mechanism to regulate the flow of information so that a fast host cannot overrun a slow one. Flow control between hosts is distinct from flow control between switches, although similar principles apply to both.

The Network Layer: Layer 3

The network layer is concerned with controlling the operation of the subnet. A key design issue is determining how packets are routed from source to destination. Routes could be based on static tables that are "wired into" the network and rarely changed. They could also be determined at the start of each conversation—for example, a terminal session. Finally, they could be highly dynamic, being determined anew for each packet, to reflect the current network load.

If too many packets are present in the subnet at the same time, they will get in each other's way, forming bottlenecks. The control of such congestion also belongs to the network layer.

Since the operators of the subnet may well expect remuneration for their efforts, there is often some accounting function built into the network layer. At the very least, the software must count how many packets, characters, or bits each customer sends, to produce billing information. When a packet crosses a national border, with different rates on each side, the accounting can become complicated.

When a packet has to travel from one network to another to get to its destination, many problems can arise. The addressing used by the second network may be different from the first one. The second one may not accept the packet at all because it is too large. The protocols may differ, and so on. It is up to the network layer to overcome all these problems to allow heterogeneous networks to be interconnected.

In broadcast networks, the routing problem is simple, so the network layer is often thin or even nonexistent.

The Data-Link Layer: Layer 2

The main task of the data-link layer is to take a raw transmission facility and transform it into a line that appears free of transmission errors in the network layer. It accomplishes this task by having the sender break up the input data into data frames (typically a few hundred bytes), transmit the frames sequentially, and process the acknowledgment frames sent back by the receiver. Since the physical layer merely accepts and transmits a stream of bits without any regard to meaning of structure, it is up to the data-link layer to create and recognize frame boundaries. This can be accomplished by attaching special bit patterns to the beginning and end of the frame. If there is a chance that these bit patterns might occur in the data, special care must be taken to avoid confusion.

The data-link layer should provide error control between adjacent nodes.

Another issue that arises in the data-link layer (and most of the higher layers as well) is how to keep a fast transmitter from drowning a slow receiver in data. Some traffic regulation mechanism must be employed in order to let the transmitter know how much buffer space the receiver has at the moment. Frequently, flow regulation and error handling are integrated for convenience.

If the line can be used to transmit data in both directions, this introduces a new complication that the data-link layer software must deal with. The acknowledgment frames for A to B traffic compete for the use of the line with the data frames for the B to A traffic. A clever solution (piggybacking) has been devised.

The Physical Layer: Layer 1

The physical layer is concerned with transmitting raw bits over a communication channel. The design issues have to do with making sure that when one side sends a 1 bit, it is received by the other side as a 1 bit, not as a 0 bit. Typical questions here are how many volts should be used to represent a 1 and how many for a 0, how many microseconds a bit lasts, whether transmission may proceed simultaneously in both directions, how the initial connection is established and how it is torn down when both sides are finished, and how many pins the network connector has and what each pin is used for. The design issues here deal largely with mechanical, electrical, and procedural interfaces, and the physical transmission medium, which lies below the physical layer. Physical layer design can properly be considered to be within the domain of the electrical engineer.

IEEE's View of the ISO/OSI Reference Model

The Institute of Electrical and Electronics Engineers (IEEE) has subdivided both the data-link layer and the physical layer into sublayers to attain a higher level of interoperability between devices (Figure 1.11).

The data-link layer is divided into logical link control (LLC) and the media access control (MAC) layer. LLC interfaces with the network layer and interprets commands and performs error recovery. It provides a common protocol between the MAC and network layer. The MAC layer controls the data transfer to and from the physical layer.

The physical layer is subdivided into the physical layer convergence procedure (PLCP) and the physical medium dependent (PMD).

Bottom Most Layer of OSI Reference Model

IEEE's subdivision of the two bottom layers of OSI Reference Model

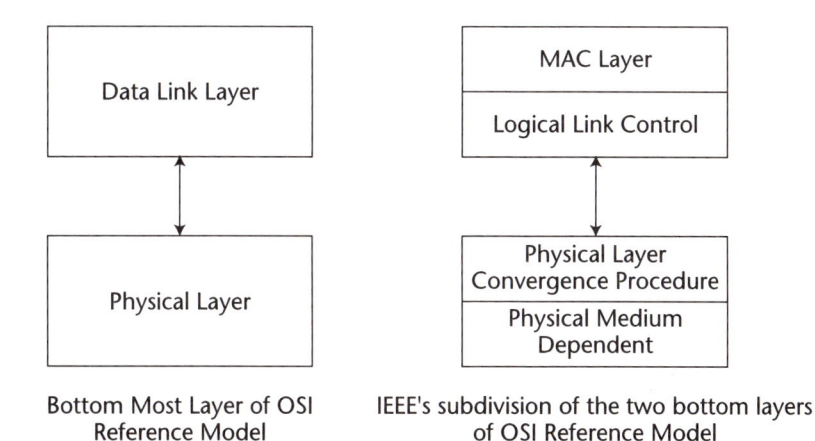

Figure 1.11 IEEE's ISO/OSI subdivision.

PLCP properly maps the MAC-specified data to the format that can be understood by the PMD layer and vice versa. The PMD layer provides the point-to-point communications between computers in the network. For example, on an Ethernet network, PMD on the network card communicates with PMDs of other network cards to establish communication between the computers.

IEEE's subdivision has enabled both software and hardware vendors to develop solutions that interoperate with each other and are easier to implement.

Network Interface Cards (NIC)

Hardware network adapters implement the physical layer of the OSI layer. Almost all computers today use one of the IEEE standard cards to add the networking functionality. The NICs are technically named after the IEEE standard that they follow along with the physical connectivity and type of media they use. For example, an Ethernet NIC works with a MAC adapter that knows how to format data for the IEEE 802.3 Ethernet standard. A twisted pair Ethernet adapter connects to the network with a twisted pair cable and follows the IEEE Ethernet standard. Commonly used network adapters include Ethernet NICs and Token Ring NICs.

Networking Cable and Physical Connections

In all wired networks, an NIC is connected with the network through NIC-supported connectors and cables. There are two major types of cables used with LANs, these are twisted pair cable and coaxial cable.

Twisted Pair Cable

Twisted pair cables (see Figure 1.12) are available both as shielded and unshielded. The cable has four pairs of wires inside the jacket. Each pair of wires is twisted with a different number of twists per inch to help eliminate interference from adjacent pairs and other electrical devices.

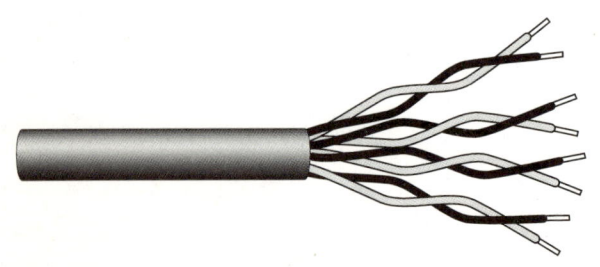

Figure 1.12 Twisted pair cable.

The tighter the cable is twisted, the higher the supported transmission rate and the greater the cost per foot. The Electronic Industry Association/Telecommunication Industry Association (EIA/TIA) have established standards for unshielded twisted pair (UTP) cables. There are five categories of UTP cables (see Table 1.1).

When selecting the network cable, you should choose the best cable you can afford. This helps in upgrading the network in the future when faster technologies are available.

Unshielded twisted pair cables have the disadvantage of being susceptible to radio and electrical frequency interference. Shielded twisted pair is suitable for environments with electrical interference; however, the extra shielding can make the cables quite bulky. Shielded twisted pair is often used on networks using Token Ring topology.

Coaxial Cable

Coaxial cabling (see Figure 1.13) has a single copper conductor at its center. A plastic layer provides insulation between the center conductor and a braided metal shield. The metal shield helps to block any outside interference from fluorescent lights, motors, and other computers.

Although coaxial cabling is difficult to install, it is highly resistant to signal interference. In addition, it can support greater cable lengths between network devices than twisted pair cable. The two types of coaxial cabling are thick coaxial and thin coaxial.

Thin coaxial cable is also referred to as thinnet. 10Base2 refers to the specifications for thin coaxial cable carrying Ethernet signals. The 2 in 10Base2 refers to the approximate maximum segment length, which is 200 meters. In actuality, the maximum segment length is 185 meters. Thin coaxial cable is popular in school networks, especially linear bus networks.

Table 1.1 The Five Twisted Pair Cable Categories

CATEGORY	USE
1	Voice Only (Telephone Wire)
2	Data up to 4 Mbps (LocalTalk)
3	Data up to 10 Mbps (Ethernet)
4	Data up to 20 Mbps (16 Mbps Token Ring)
5	Data up to 100 Mbps (Fast Ethernet)

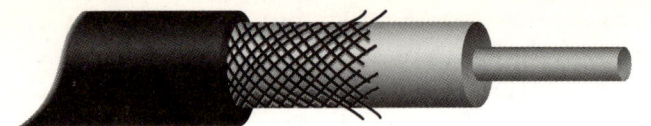

Figure 1.13 Coaxial cable.

Thick coaxial cable is also referred to as thicknet. 10Base5 refers to the specifications for thick coaxial cable carrying Ethernet signals. The 5 in 10Base5 refers to the maximum segment length being 500 meters. Thick coaxial cable has an extra protective plastic cover that helps keep moisture away from the center conductor. This makes thick coaxial a great choice when running longer lengths in a linear bus network. One disadvantage of thick coaxial is that it does not bend easily and is difficult to install.

Hubs

Hubs are used in situations where two or more computers need to be physically wired together (see Figure 1.14). In other words, hubs physically connect computers on a LAN.

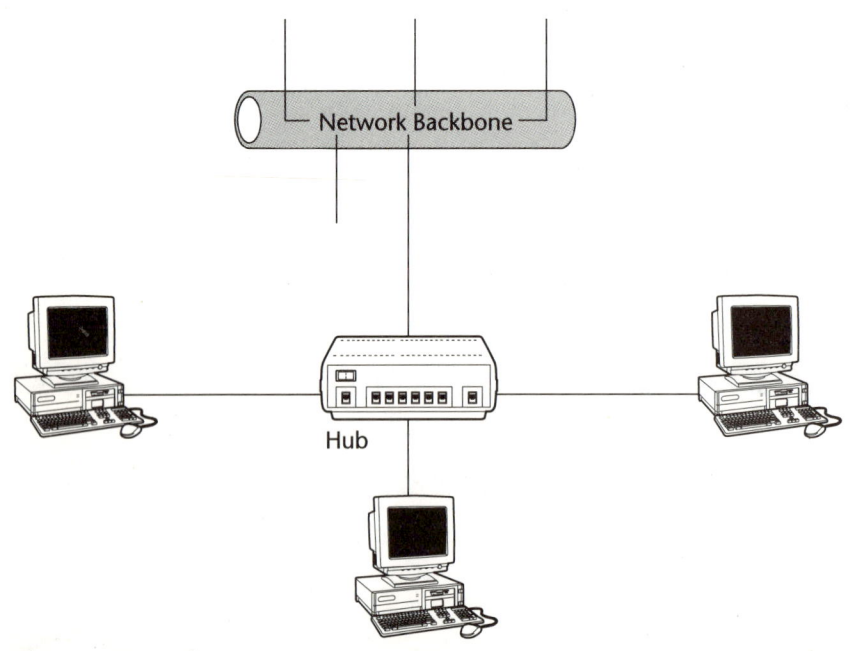

Figure 1.14 Hub.

Hubs can be chained together to extend the number of computers participating on a network.

Routers

Routers (see Figure 1.15) restrict and route the network data traffic on a network. Consider a scenario where two different departments are interconnected with each other using the same network; assume that the two departments hardly need to communicate with each other. Because they both share the same network bandwidth, the networks get jammed and a little too busy. But if the network is divided into two separate networks and a router is put in between them, then the network is much cleaner and does not get clogged or too busy, as each department is concerned only with its own traffic and does not have to be concerned with the other's. Whenever data needs to be sent to the other department, the router acts as a network traffic controller and simply allows that data to pass through to the other network.

Routers, therefore, simplify the network and greatly improve the network performance.

Repeaters

Wired LANs can cover a limited geographical area, which usually ranges from 150 to about 300 meters with most wired networks. The maximum range that a LAN can cover depends upon the equipment and the type of cable used. Repeaters are a simple solution to overcome and extend the geographic limit. The reason for the limited area that LANs cover lies in the fact that electrical signals become weaker as they travel on a medium. Repeaters are devices that act like a relay station and strengthen an incoming weak electrical signal, without any alteration in the data that signal carries, and retransmit the data for further use. Repeaters should be placed at distances whenever a weaker signal is detected. In most networks, repeaters are needed at every 150 to 300 meters.

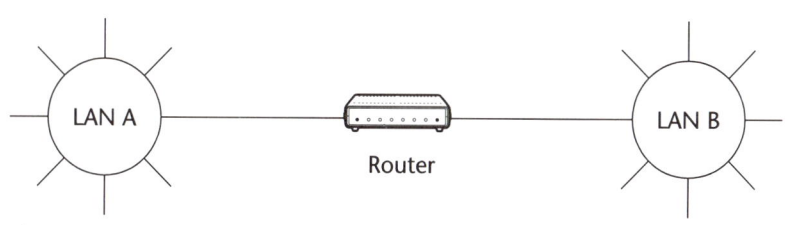

Figure 1.15 Router.

Networking Software

In order to access a network, a user must install network software on his or her computer. The network software includes the proper network protocols and the NIC drivers.

A common example of the application software one might want to use would be a Web browser. A Web browser uses the network software to communicate with another computer and displays the results of the communication.

The networking protocols identify the computer and the user on a network to another computer and user. The most widely used network protocol is Transmission Control Protocol/Internet Protocol, or TCP/IP (see the next section on TCP/IP), which is also used on the Internet.

NIC drivers are normally devised by the NIC manufacturer and are set according to their specifications. Network drivers must be made compatible with the operating system. Network drivers communicate both with the networking protocols and the LLC to facilitate the data transmission over the wire.

Networking Protocol: TCP/IP

Networking protocols provide computer application software to access the network. These protocols provide an abstraction of the computer hardware, operating system, and physical characteristics of the network.

As already mentioned, TCP/IP is by far the most commonly used protocol, so its basic operation bears some examination. Many of the overall principles used in this protocol apply to other types of protocols. As a result of the explosive growth that the Internet has seen over the past decade, TCP/IP has become the de facto standard protocol for networking. Most vendors have dropped their proprietary protocols and adopted TCP/IP as the protocol for their networking software. (WAP is irrelevant for this discussion. This book is about 802.11b, which is essentially wireless Ethernet.)

The history of TCP/IP and the Internet begins in 1973, when the U.S. DoD Advanced Research Projects Agency (DARPA) initiated a research program to investigate techniques and technologies for interlinking packet networks of various kinds. The objective was to develop communication protocols that would allow networked computers to communicate transparently across multiple, linked packet switching networks. The network was initially known as ARPANET. One of the lasting legacies of ARPANET was a host of protocols that worked on packet switching network protocols including TCP/IP. The system of interconnected networks that emerged from this research eventually became commonly known as the Internet. The initial network protocol adapted by ARPANET was known as Network Control Protocol (NCP). By 1974 NCP was deemed inadequate to handle the growing traffic over the rapidly expanding network. At that time a more robust network Transmission

Control Protocol (TCP) was adopted. The initial TCP design defined both the information required for the routing of the data-packets from one end to the other as well as structure of the data or payload. This protocol was considered too heavyweight for the intermediate routers because they had to deal with end-to-end data. So in 1978 this protocol was divided into parts: one to handle the routing of data-packets, the other to handle end-to-end data transmission. The system of protocols that was developed over the course of this research effort became known as the TCP/IP Protocol Suite, after the two initial protocols developed: Transmission Control Protocol (TCP) and Internet Protocol (IP). TCP corresponds to the transport layer of the ISO/OSI model, and IP is the implementation of the network layer.

The current version of IP is IPv4, and the upcoming version is IPv6. IPv4 or the current implementation of IP that is used throughout the Internet uses 32-bit addresses commonly represented by a set of four 8-bit numbers ranging from 0 to 255 separated by periods or dots. This is commonly known as the IP address. Each IP address identifies a particular node in the network. With the growth of the Internet this address space is rapidly being depleted and there is need for a wider address space. As a response to this demand, IPv6, which has a 32-bit address space, has been developed.

The basic parameters of a TCP/IP network include IP address, subnet mask, Internet naming and domain name servers, default gateway, and IP routing. The next sections discuss each of these.

IP Address

Each computer participating on a TCP/IP network must have a unique IP address. An IP address in an IPv4 is a 32-bit number represented as a set of four bytes, with each number ranging from 0 to 255. The IP address is normally represented as set of four numbers separated by a period. This format is known as the dotted decimal format. For example, 192.168.0.2 is an IP address in the dotted decimal format.

For a computer to participate on the Internet it must have a unique IP address. The IP addresses to be used on the Internet were originally assigned by the Internet Network Information Center, or InterNIC, which was operated on behalf of the National Science Foundation (NSF) by Network Solutions Inc. (NSI). NSI was formed under a five-year contract granted in 1993 to assign Internet names and addresses and educate the general public about the Internet. Since April 1998, the IP address space, and all TCP/IP-related numbers, has historically been managed by the Internet Assigned Numbers Authority (IANA), a nonprofit industry organization (www.iana.org), under the auspices of the U.S. Department of Commerce, which now holds the authority over the Internet. (For more information, go to www.internic.net.) IANA generally allocates IP addresses to the service providers and large organizations. When the IP address scheme was initially proposed, the IP address space was divided into three classes of addresses used in IP-based networks. These classes were known as class A, B, and C. Each was intended for use with a different size of network, with each class A network capable of having 16,581,373 addresses or 1/255 of the total address space. There were also some addresses set aside for those networks that were not connected directly to the Internet. These are also known as unrouted networks. With the explosive growth of the Internet, this address allocation scheme resulted in a severe shortage of addresses for the newcomers and excessive waste for the ones who had registered addresses earlier. This resulted in the reallocation of address space and a new system of managing the addresses. IANA now allocates addresses only to very large organizations and service providers, who in turn allocate addresses to their subscribers in their address space. Typically service providers provide their subscribers with a small set of addresses that are typically assigned to the routers and firewalls connected directly to the Internet. Most of the computers inside a private network use the unrouted address internally and connect to the Internet through firewalls. There is also an address class called class D that is reserved for multicasting; for our purposes, we do not need to know about this type.

Class A addresses (see Figure 1.16) are intended for use with networks that have a large number of attached hosts (up to 2^{24}); class C addresses allow for a large number of networks each with a small number of attached hosts (up to 256). An example of a class A network is ARPANET, and an example of a class C network is a single LAN. Class A network addresses have 7 bits for the network identifier or netid and 24 bits for the host identifier or hostid; class B addresses have 14 bits for the netid and 16 bits for the hostid; and class C addresses have 21 bits for the netid and 8 bits for the hostid. 10.1.1.1 is an example of a class A IP address. In this network the netid is 10. 16.72.0.3 is an example of a class B IP address; here the netid is 16.72. 192.1.1.2 is an example of a class C IP address; here the netid is 192.1.1.

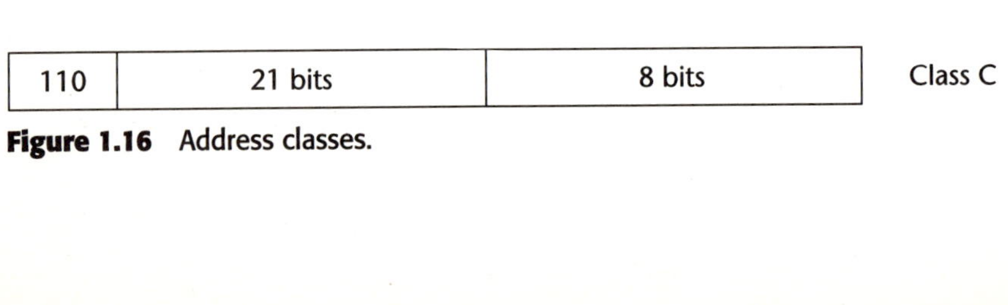

Class id	netid	hostid	
0	7 bits	24 bits	Class A

Class id	netid	hostid	
10	14 bits	16 bits	Class B

Class id	netid	hostid	
110	21 bits	8 bits	Class C

Figure 1.16 Address classes.

Subnet Mask

Subnet masks are used to efficiently utilize the IP addresses within a LAN. Address masks consist of binary 1s in the positions that contain the network address and binary 0s in the positions that contain the hostid. The routers remember the subnet mask. All IP packets are routed based on the IP address and the subnet mask. For example, a subnet mask of 255.255.255.0, when combined with an IP address of 192.168.0.2, helps the router to properly route the IP packet.

Internet Naming and Domain Name Servers

It can be extremely difficult to remember all the numerical host IP addresses with which you might want to communicate. Instead, TCP/IP supports host naming, which allows an Internet name to be associated with an IP address; these names are called host names or the domain host names. For example, www.wiley.com is the domain host name of this book's publisher, John Wiley and Sons—the name corresponds to the numerical address, and both are stored in databases called domain name servers. TCP/IP includes special support for looking up IP addresses for the host names and vice versa. In order to correctly address a computer using a host name, a valid DNS must be configured under the IP settings on a computer, and those DNS servers must be available and accessible when IP address lookup is desired.

Default Gateway

Default gateway is the term used for identifying the router available on a network. All local LAN traffic must go through the router to reach another part of the LAN or the Internet. IP packets originating at a LAN are received by the gateway, and the gateway properly routes the IP address to the intended LAN.

IP Routing

With a limited number of IP addresses available, it was important to use a scheme to efficiently utilize the IP address pool. Routers with subnet masks are used to accomplish this purpose. Routers are used to separate logical networks and are assigned the netid as their IP addresses. All the hosts, which reside inside the router's domain, are required to use the IP address from the same netid host pool.

Putting It All Together

Now that we know most of the basic things about a LAN, let's try to step through a simple LAN setup. These steps help you understand the normal process you should follow when setting up a new LAN.

Let's assume that we are making a LAN consisting of four Microsoft Windows-based computers and we want to use Ethernet LAN adapters with twisted pair cabling.

The steps for installing a network are as follows:

1. Install the network adapters in the computers according to the vendor instructions.
2. Connect the computers together using an Ethernet hub and the twisted pair network cable.
3. Turn on all the computers and install the network adapter drivers per vendor's instructions.
4. For TCP/IP settings, assign IP addresses to each of the computers starting with 192.168.0.2 through 192.168.0.5, with the subnet mask on each computer set to 255.255.255.0.
5. Reboot the machines according to the operating system and the vendor's instructions.
6. When the computers have rebooted, go to the command prompt (on Windows Platform, run either command.exe or CMD.exe) and type "ping 192.168.0.2" from the computer with IP address 192.198.0.3. The program should reply with:

```
C:\>ping 192.168.0.2
Pinging 192.168.0.2 with 32 bytes of data:

Reply from 192.168.0.2: bytes=32 time<10ms TTL=128
Reply from 192.168.0.2: bytes=32 time<10ms TTL=128
Reply from 192.168.0.2: bytes=32 time<10ms TTL=128
Reply from 192.168.0.2: bytes=32 time<10ms TTL=128
```

If there is an error or the computer is not configured correctly, you might get an error as follows:

```
C:\>ping 192.168.0.2
Pinging 192.168.0.2 with 32 bytes of data:
Request timed out.
Request timed out.
Request timed out.
Request timed out.
```

If you get the error message, check the operating system instructions and the instructions provided by the network card provider. We discuss setting up and troubleshooting the wireless LANs in the chapters to come.

Summary

In this chapter, we explored the basics of a computer network. We explained the history of the computer networks, the basic topologies, the protocols, the network cards, and the network hardware that a LAN consists of. Finally, we set up a hypothetical wired LAN to ensure that we understand the basic concepts. In the next chapter, we cover wireless LANs, from the basics to the architecture of a wireless LAN.

CHAPTER 2

Wireless LANs

Wireless communications have enjoyed a steady growth over the last couple of decades. From television remote control to satellite-based communication systems, wireless communications have changed the way we live. Devices connected through wireless technology provide increased mobility and require less infrastructure than traditional wired networks. Computer networks have lagged behind in the wireless race because of intrinsic needs for higher bandwidth for data transmission compared to that of existing wireless devices (for example, television remote control or cordless phones). However, in recent years, breakthroughs in silicon-chip technology have increased data throughput over the wireless connections, making wireless computer networks a reality. Using electromagnetic waves, wireless LANs transmit and receive data over the air, minimizing the need for wired connections. With today's technology, wireless LANs are highly scalable, reliable, and easy to implement.

Wireless LANs have gained significant popularity among mobile users and those who work in small groups. Wireless LANs enable mobile users to gain access to real-time information. A wireless LAN can be implemented as a standalone network (that is, a LAN with computers connected only using wireless links), with a handful of computers, as an enterprise-scale network with thousands of computers, as an extension to an existing wired network, or as a replacement to an existing wired network.

In this chapter we present a brief evolution of wireless networks and explore the basics of wireless networks. First to be discussed are the basic components of a wireless network. Next, wireless LAN architecture and the technologies that constitute a wireless LAN are examined in detail. Then wireless networks are compared with wired networks. Finally we cover the existing standards in wireless LAN technology.

Evolution of Wireless LANs: An Overview

The U.S. Army first used radio signals for data transmission during Word War II more than 50 years ago. The army developed a radio data transmission technology, SIGSALY, which was heavily encrypted. The mere existence of the capabilities to conduct secure wireless communications was kept classified until 1976. The army filed close to 80 patents, but these were also kept secret. These were used quite extensively throughout the campaign by the United States and its allies. As the 1970s approached, computer capabilities were becoming cheaper and spreading rapidly in academic institutions. The scientists working with these computers saw that, to enable them to share their research data, their computers needed to be able to communicate with each other. Around this same time ARPANET was slowly adding more nodes to its network. This technology inspired a group of researchers in 1971 at the University of Hawaii to connect with ARPANET; unfortunately or fortunately the geography of Hawaiian Islands presented a challenge for connecting the computers, since this networking required wired connections, which was a monumental task considering that some of these nodes were on different islands. To overcome this challenge, they created the first packet-based radio communications network. ALOHAnet, as it was named, was essentially the very first wireless LAN. With this, wireless networking was born. This first wireless LAN consisted of seven computers on four islands communicating with the central computer on the Oahu Island in a bidirectional star topology. A bidirectional star topology configuration consists of systems that are connected to a central system known as a hub, and they can send and receive data at the same time.

In Europe, a Swedish scientist named Östen Mäkitalo, also known as "Mr. Mobile," is considered the brain behind the first wireless network. Östen was working as the research and development director of Sweden's national telephone company, Televerket (Telia), when he was asked to develop a common system for connecting the mobile phones in all of the Scandinavian countries.

He brought in research teams from Denmark, Norway, and Finland and created the Nordic Mobile Telephone (NMT) system, which was launched in 1981.

As personal computers became more pervasive during the 1980s and 1990s, the demand to connect them wirelessly grew. Initially, vendors such as IBM, Digital Equipment Corporation (DEC), and Symbol Technologies offered proprietary solutions to their customers. Almost all the vendors were under pressure to create more interoperable technologies.

In 1997, the Institute of Electrical and Electronics Engineers (IEEE) drafted the 802.11 standard for wireless local area networking. The basic concepts of 802.11 were based upon Ethernet, which by this time had become the de facto standard for wired LANs. The initial 802.11 standard had left several key questions such as the encoding schemes up to each vendor's discretion, which resulted in the development of incompatible equipment. 802.11 was also limited to a 2-megabits-per-second transmission rate. During this time several advances in direct-sequence spread spectrum (DSSS) technology and relaxation of FCC rules allowed the IEEE to draft the 802.11b standard in 1999, which was accepted by the networking industry, and products for wireless networking over the 2.4-GHz frequency entered the market. The details of both of these standards are covered later in this chapter.

Initially, 802.11 standard-based wireless LANs operated at 1 megabit per second (Mbps). Eventually, 802.11b—also known as wireless fidelity, or Wi-Fi, or 802.11 high rate, which is a modified version of the 802.11 standard—operated at 11 Mbps. Today, higher speeds are being achieved through newer standards, optimized usage of wireless communication methods, and improvement in computer hardware. Fortunately, the wireless LAN industry has matured enough to accept the need of standards, and IEEE is doing a great job of establishing standards in the wireless arena.

A Basic Wireless LAN

A simple wireless LAN (see Figure 2.1) consists of two or more computers connected via a wireless link to communicate with each other. A link on a wireless LAN does not consist of a cable or any type of physical connection; instead it consists of a connection via electromagnetic spectra traveling over the air or ether in which data is transmitted. The computers in a wireless network require network interface cards (network adapters) that establish and maintain the transmission and reception of information between the networked computers. Also, each computer on a wireless network must conform to the same technological standard (that is, use the same networking technology).

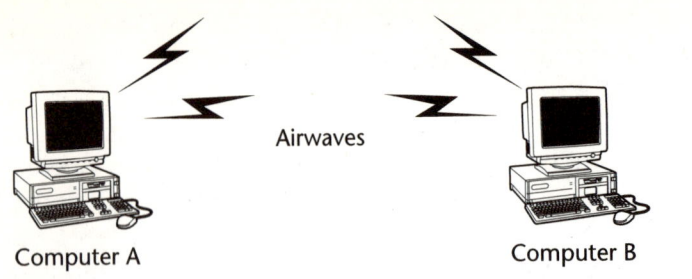

Figure 2.1 Two computers interconnected over a wireless network.

All computers in a wireless LAN must have appropriate computer hardware that enables wireless connectivity. Wireless LAN hardware is an electronic component that is attached to a computer that needs to be connected to a wireless LAN. These electronic components are known as wireless LAN adapters or wireless LAN network interface cards (NICs).

The network adapters can be implemented as Personal Computer Memory Card International Association (PCMCIA) cards in notebook computers, Industry Standard Architecture (ISA) or Peripheral Component Interconnect (PCI) adapters in desktop computers, and are often fully integrated devices within handheld computers. Each of these network adapters must contain the necessary transceivers (devices that perform both transmission and reception) for the media of choice. Similar to an FM radio where the sound is superimposed on the radio wave, in wireless LANs the data to be transmitted is superimposed on the electromagnetic carrier at the specified frequency and transmitted over the airwaves by the wireless network adapter (see Figure 2.2). The network adapter at the receiving end listens at the same frequency, and when it receives the transmitted waves, it extracts the superimposed data from the electromagnetic carrier and pushes up the technology stack for usage.

For a wireless network to be successful, it must be reliable, safe, secure, fast, and easy to deploy. In the next section, we describe the architecture of a wireless LAN.

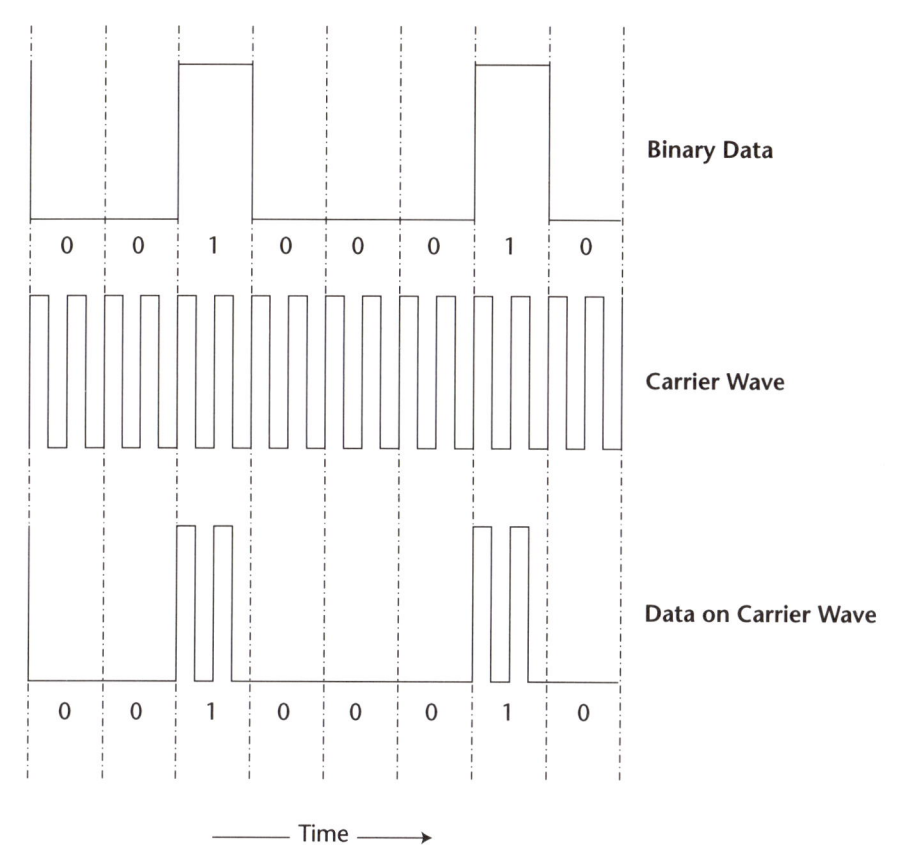

Figure 2.2 Data transmission over carrier waves.

Basic Architecture of a Wireless LAN

A wireless LAN can be deployed in many ways, depending on the architecture it is built on. There are many wireless devices available today, so it is best to choose devices that interoperate when building a wireless LAN. This section explains wireless LAN hardware, including wireless network adapters and access points. We discuss their configuration and the standards they follow.

Wireless LAN Adapters

Each and every computer in a wireless LAN must have a wireless network adapter. Having this device in place and in operation ensures that the different computers' OSI models (see Chapter 1, "Networking Basics," for more information on the OSI Reference Model) will be compatible with one another.

Figure 2.3 shows an OSI Reference Model for a generic wireless LAN. For our concerns with wireless networks, we need only reexamine the bottom two layers of the model—the physical layer and the data-link layer—as there is some variation here between the wireless and nonwired networks. For any given implementation the higher layers do not differ appreciably between the two kinds of networks. Wireless LAN adapters (network interface cards) implement the physical layer of the OSI Model and conform to the data-link layer to provide proper media access control (MAC)-level interface.

The Physical Layer

The primary difference between a wireless LAN adapter and a wired LAN adapter is its physical layer. The physical layer in wireless adapters does two things: When data is transmitted, it converts the electronic signals into medium-dependent signals, and when signals are received from the medium, it converts them into electronic signals for higher layers to interpret. How well a wireless network is going to perform depends on the bandwidth of the electromagnetic spectrum, the quality of the components that make up the network adapter, and the protocols used.

For electrical signals (such as those used in cordless phones, FM radio, and wireless LANs) to propagate through the air, they must be converted into electromagnetic waves. An electromagnetic wave is energy that travels and spreads out as it goes—for example, visible light that comes from a lamp in your house or radio waves that come from a radio station. The distance that an electromagnetic wave can travel depends on its wave properties. These properties include the wavelength and the frequency of an electromagnetic wave. Scientists and industry standard committees have divided and named different types of electromagnetic waves into bands based on their properties (see Figure 2.4). Each electromagnetic band has a certain minimum frequency and a maximum frequency, and all the radiation that falls between the two is known as the band frequency. The entire set of known electromagnetic waves is called the electromagnetic spectrum. Examples of electromagnetic waves are radio frequencies (RF), microwaves, infrared and ultraviolet light, X-rays, and gamma rays. Due to the highly popular nature of wireless connectivity and the limited availability of the electromagnetic spectrum, wireless LANs use a limited bandwidth of the electromagnetic spectrum to transmit data over the ether or air.

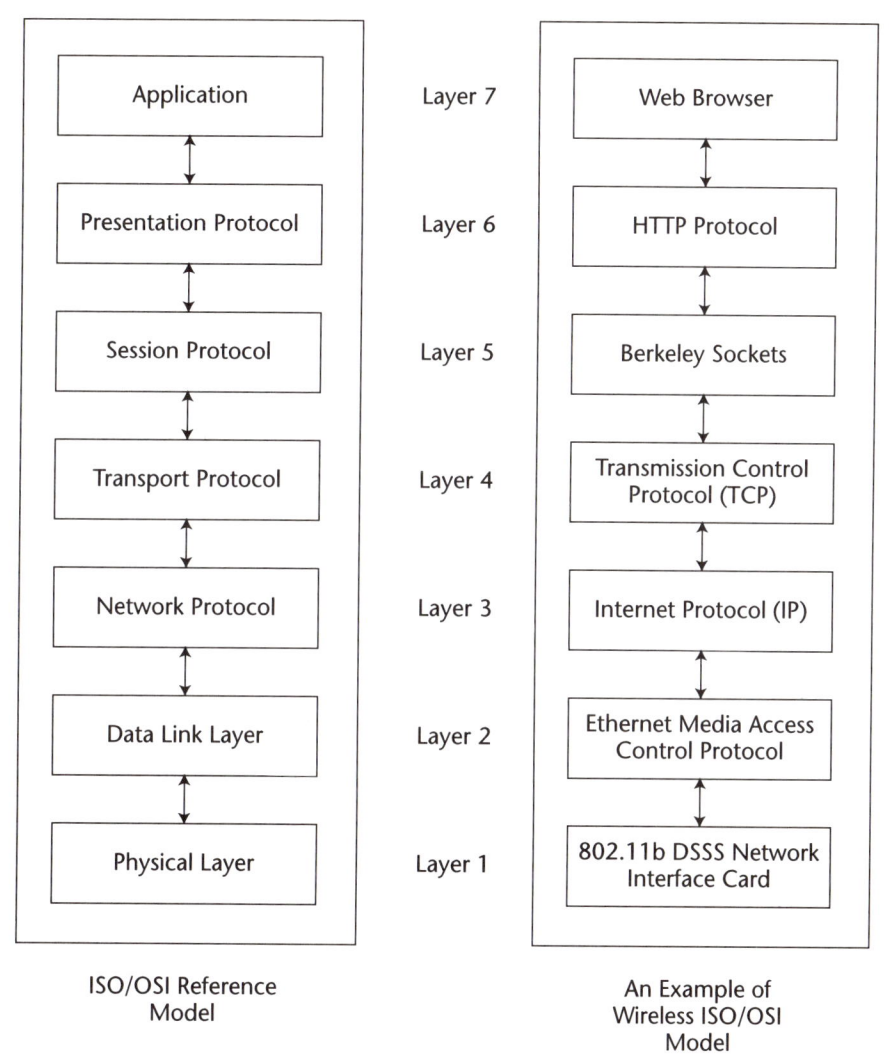

ISO/OSI Reference Model		An Example of Wireless ISO/OSI Model
Application	Layer 7	Web Browser
Presentation Protocol	Layer 6	HTTP Protocol
Session Protocol	Layer 5	Berkeley Sockets
Transport Protocol	Layer 4	Transmission Control Protocol (TCP)
Network Protocol	Layer 3	Internet Protocol (IP)
Data Link Layer	Layer 2	Ethernet Media Access Control Protocol
Physical Layer	Layer 1	802.11b DSSS Network Interface Card

Figure 2.3 OSI Reference Model for a wireless LAN adapter.

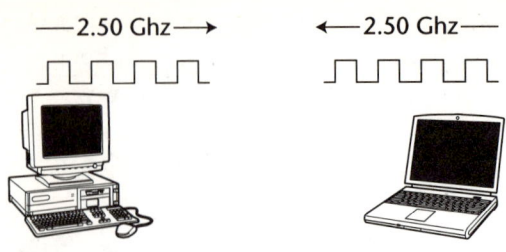

Figure 2.4 Electromagnetic spectrum.

Today, there are three basic electromagnetic spectrum bands that are commonly used for data transmission over wireless LAN links: infrared, radio frequency, and microwave.

Infrared

Infrared-based systems are the simplest and least expensive wireless LANs. These systems work best when operated in the line of sight (that is, the transceivers involved in communication must face one another without any physical obstruction). Infrared systems are not bandwidth constrained—in other words, the signal does not spread, so devices using infrared can use the entire bandwidth of infrared when communicating with one another without any interference with any other device. They also can attain high speeds at relatively low costs compared to other competing types of systems. Another benefit of using an infrared-based system is that it does not require any licensing from the Federal Communications Commission (FCC), which only regulates the RF portion of the electromagnetic radiation between 9 kHz and 300 GHz. Infrared radiation falls in the light portion of the electromagnetic spectrum that is not regulated by the federal government. A precisely aimed infrared system can attain a high range, up to several miles, which can be good for internetworking. However, when you need omni-directional connectivity, where the signals are bounced off nearby objects in all directions, system performance is reduced. Infrared systems do not perform well under such conditions. Infrared systems also suffer from interference from sunlight and artificial light. Initially, infrared systems were very popular, but their unreliability due to easily obstructed signals means these systems have limited use.

Microwave

Microwave (MW)-based networks normally operate in the 5.8-GHz band and use less than 500 milliwatts of power. The typical range for a microwave-based system in a closed office environment is about 120 feet. The big advantage to MW systems is higher throughput, as MW systems do not have the overhead involved with spread spectrum systems. MW-based adapters are normally used in commercial settings.

Radio

Radio frequency (RF)-based wireless LANs are by far the most popular wireless LANs in the United States. FCC regulations require that RF systems use spread spectrum technology (see Figure 2.5). Spread spectrum modulation techniques are defined as being those techniques in which the following occur:

1. The bandwidth of the transmitted signal is much greater than the bandwidth of the original message.
2. The bandwidth of the transmitted signal is determined by the message to be transmitted and by an additional signal known as the spreading code.

Spread spectrum was originally designed for the U.S. Navy to hide the signals that controlled torpedoes. The topic was classified for years by the U.S. military, and only recently were the patents made public. Systems that use spread spectrum must use the exact same frequency and related parameters. These parameters are defined by a particular implementation of spread spectrum technology. The commonly used RF bands in most countries for wireless LANs are 2.4 GHz and 5.x GHz (the x in 5.x varies).

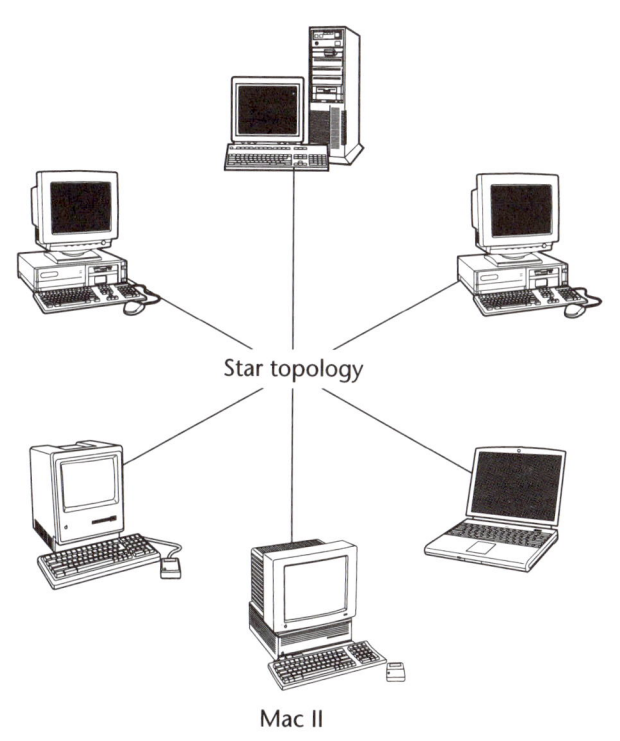

Figure 2.5 A spread spectrum network.

Currently two types of spread spectrum technology exist: direct-sequence spread spectrum (DSSS) and frequency-hopping spread spectrum (FHSS). Because the signal does not stay in one place on the band, FHSS can elude radio interference. DSSS avoids interference by configuring the spreading function in the receiver to concentrate the desired signal and spread out and dilute any interfering signal. There is a lot of overhead involved with spread spectrum, and so most of the DSSS and FHSS systems historically have lower data rates than infrared- or microwave-based systems.

Direct-sequence spread spectrum (DSSS). In the direct-sequence spread spectrum (DSSS), the transmission signal is spread over an allowed band (see Figure 2.6). The data is transmitted by first modulating a random binary string called spreading code on the carrier wave (the chosen electromagnetic frequency). The data bits are mapped to a pattern of ratios of spreading code bits called chips and mapped back into bits at the destination. Looking at it another way, the number of chips that represent a bit is the spreading ratio. The higher the spreading ratio, the more the signal is resistant to interference. The lower the spreading ratio, the more bandwidth is available to the user. The FCC mandates that the spreading ratio must be more than 10. Most products have a spreading ratio of less than 20. The transmitter and the receiver must be synchronized with the same spreading code. If orthogonal spreading codes are used, then more than one LAN can share the same band. However, because DSSS systems use wide subchannels, the number of colocated LANs is limited by the size of those subchannels. Recovery is faster in DSSS systems because of the ability to spread the signal over a wider band.

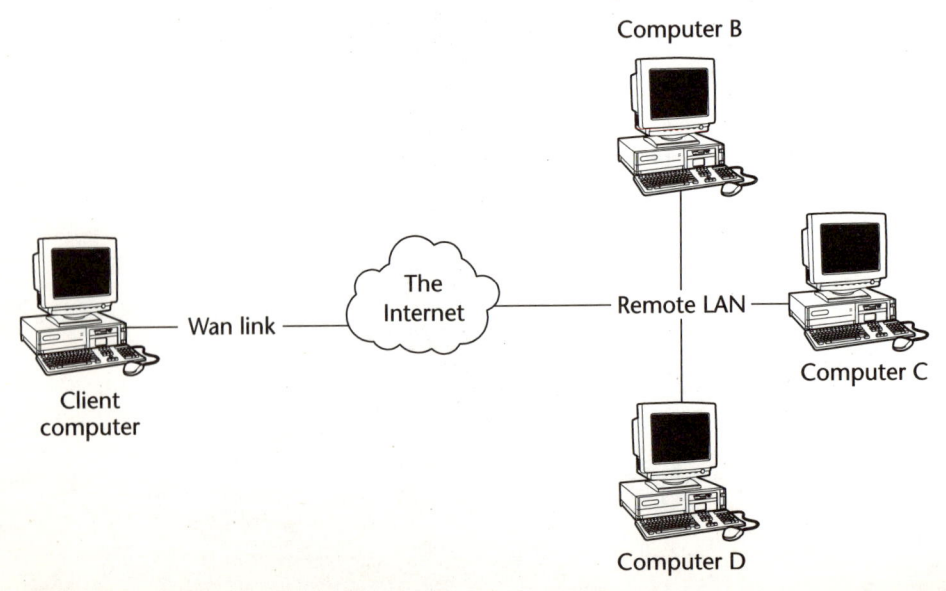

Figure 2.6 DSSS operation.

Frequency-hopping spread spectrum (FHSS). This spread spectrum technique divides the frequency band into smaller subchannels of usually 1 MHz (see Figure 2.7). For example, in most wireless LANs the 2.4-GHz band is divided into 11 1-MHz subchannels. The transmitter then hops (switches) between the subchannels, sending out short bursts of data for a given time. The maximum amount of time that a transmitter spends in a subchannel is called the *dwell time*. In order for FHSS to work correctly, both communicating ends must be synchronized (that is, both sides must use the same hopping pattern); otherwise they lose the data. FHSS is more resistant to interference because of its hopping nature. The FCC mandates that the band must be split into at least 75 subchannels and that no subchannel is occupied for more than 400 milliseconds. There is an ongoing debate about the security that this hopping feature provides. Though it is possible to monitor the hopping sequence and then wait till the whole sequence is repeated, the level of security is sufficient that expensive equipment would be required to break in. Many FHSS LANs can be colocated if an orthogonal hopping sequence is used (see *Orthogonal Frequency Division Multiplexing* for more information). Because the subchannels in FHSS are smaller than in DSSS, the number of colocated LANs can be greater with FHSS systems. The most commonly used standard based on FHSS is HomeRF.

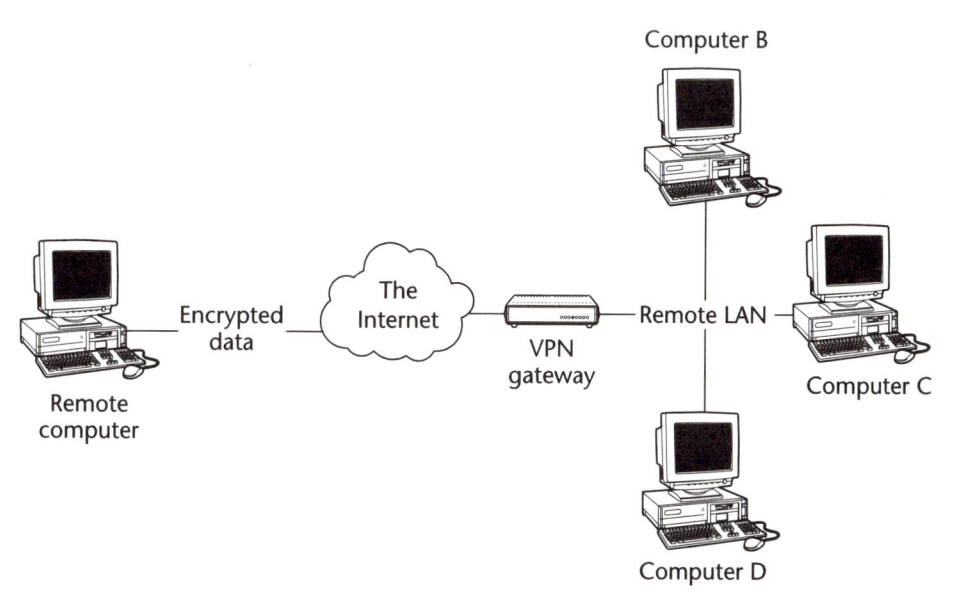

Figure 2.7 FHSS operation.

Orthogonal Frequency Division Multiplexing (OFDM). OFDM technique distributes the data to be transmitted into smaller pieces, which are simultaneously transmitted over multiple frequency channels that are spaced apart. This spacing provides the orthogonality that prevents the demodulators from seeing frequencies other than their own.

When transmitting data using the OFDM, the data is first divided into frames and a mathematical algorithm known as Fast Fourier Transformation (FFT) is applied to the frame, then OFDM parameters (for example, timing) are added. An Inverse Fast Fourier Transformation (IFFT) is then applied on each frame. The resulting frames are then transmitted over the designated frequencies. A receiver performs the inverse operations to get the transmitted data by performing FFT on the frames.

The benefits of OFDM are high spectral efficiency, resiliency to RF interference, and lower multipath distortion.

In summary, wireless LAN adapters can use infrared, microwave, or radio frequency as transmission band. Wireless LAN cards used in a given wireless network must use the same band. Wireless LAN adapters communicate with the physical media on the physical side of the reference model, and the MAC layer at the data-link layer.

Data-Link Layer

The MAC layer of the data-link layer (see Figure 2.8) controls how data is to be distributed over the physical medium. The main job of the MAC protocol is to regulate the usage of the medium, and this is done through a channel access mechanism. A channel access mechanism is a way to divide the available bandwidth resource between subchannels, the radio channel, by regulating the use of it. It tells each subchannel when it can transmit and when it is expected to receive data. The channel access mechanism is the core of the MAC protocol. With more companies and individuals requiring portable and mobile computing, the need for wireless local area networks continues to rise throughout the world. Because of this growth, IEEE formed a working group to develop a media access control (MAC) and physical layer (PHY) standard for wireless connectivity for stationary, portable, and mobile computers within a local area.

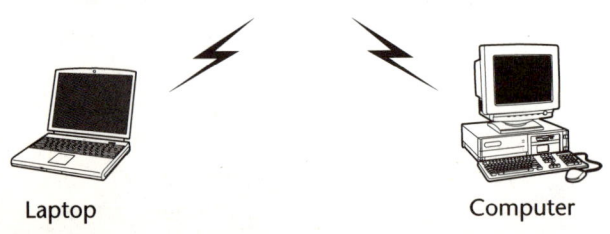

Laptop   Computer

Figure 2.8 Wireless MAC layer.

As most wired LAN products use Carrier Sense Multiple Access with Collision Detection (CSMA/CD, also known as Ethernet) as the MAC protocol, it has been a logical choice for most wireless LAN equipment manufacturers and the standard bodies to incorporate the CSMA/CD or a similar protocol as the MAC protocol. Most data-link layer implementations use either CSMA/CD or one of its derivatives, for example, Carrier Sense Multiple Access with Collision Avoidance (CSMA/CA). Carrier sense means that the station will listen before it transmits. If there is already someone transmitting, then the station waits and tries again later. If no one is transmitting, then the station goes ahead and sends what it has. If two stations send at the same time, the transmissions will collide and the information will be lost. This is where collision detection comes into play. The station will listen to ensure that its transmission made it to the destination without collisions. If a collision occurs, then the stations wait and try again later. The time the station waits is determined by using a back-off algorithm, which is essentially a structured mechanism for increasing the wait time. This technique works great for wired LANs, but wireless topologies can create a problem for CSMA/CD. This problem is known as the hidden-node problem.

The hidden-node problem is shown in Figure 2.9. Node C cannot hear node A. So if node A is transmitting, node C will not know and may transmit as well. This will result in collisions. The solution to this problem is CSMA/CA. CSMA/CA works as follows: The station listens before it sends. If someone is already transmitting, it waits for a random period and tries again. If no one is transmitting, then it sends a short message. This message is called the ready-to-send (RTS) message. This message contains the destination address and the duration of the transmission. Other stations now know that they must wait that long before they can transmit. The destination then sends a short message, which is the clear-to-send (CTS) message. This message tells the source that it can send without fear of collisions. Each packet is acknowledged, which means that upon receipt the receiver must send an acknowledgment (ACK) packet. If an ACK is not received, the MAC layer retransmits the data. This entire sequence is called the four-way handshake.

Access Points (APs)

An access point (AP) is a centralized wireless device that normally does not have any computer physically attached to it. The AP controls the traffic in the wireless medium. All traffic between the communicating computers must go through the access point. Access points are often connected to the wired LANs (for example, corporate or home networks) that are usually connected to a wide area network (WAN)—for example, the Internet using a high-speed connection such as broadband (see Figure 2.10). Access points are set up in this way, to route the traffic between the wireless LAN and the network that it connects with.

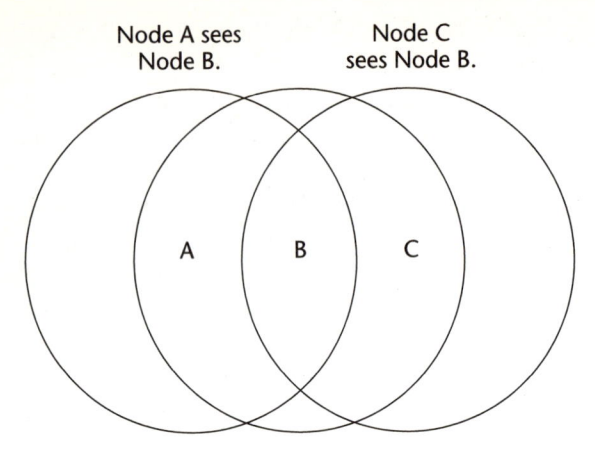

Figure 2.9 CSMA/CD hidden-node problem.

APs contain a wireless interface adapter just like any other computer in a wireless LAN. In addition, APs maintain a lot more information about the computers in the network and perform the necessary authentication, encryption, and session (connection) management with all the connected computers. If an AP is able to connect the wireless LAN with any other type of network, then it must also act as a router. In this case, it also contains a regular network interface card to participate in the other wired network.

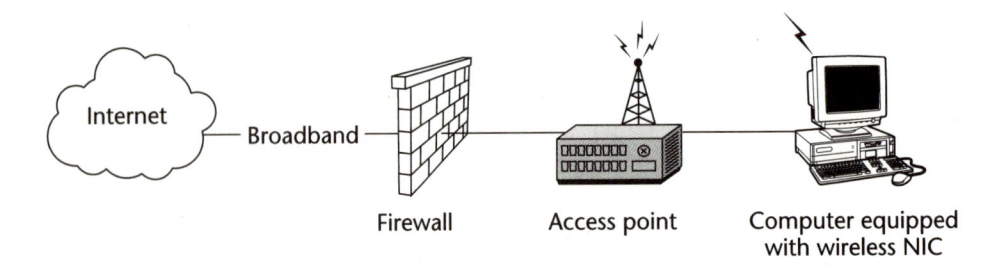

Figure 2.10 Access point connected with a broadband connection.

Wireless LAN Configurations

Each computer in a wireless LAN is referred to as a station (STA). A wireless LAN can be set up as a peer-to-peer network (called an ad-hoc network) where two or more stations directly talk to each other, or in the infrastructure mode where a central AP is involved and all communication between the stations is routed through the central AP.

Ad-Hoc Mode

When two or more stations come together to communicate with each other, they form a basic service set (BSS). The minimum BSS consists of two stations. A BSS that stands alone and is not connected to an AP is called an independent basic service set (IBSS) or an ad-hoc network (see Figure 2.11). An extended service set (ESS) is formed when two or more BSSs operate within the same network. An ad-hoc network is a network in which stations communicate only peer-to-peer. There are no APs, and no one gives permission to talk. Mostly these networks are spontaneous and can be set up rapidly.

Ad-hoc mode is rarely used, and, when set up, it is only used for temporary purposes.

Infrastructure Mode

A wireless LAN is said to be operating in infrastructure mode (see Figure 2.12) when two or more BSSs are interconnected using an AP. APs act like hubs for wireless stations. An AP routes the traffic between the two BSSs. An AP is sometimes connected to a wired network to provide wired network resources to the wireless stations. Each BSS becomes a component of an extended, larger network. An AP is a station, thus addressable as a router or gateway and routes the network traffic that is intended for the wired network and vice versa. So data moves between the BSS and the wired network with the help of these APs.

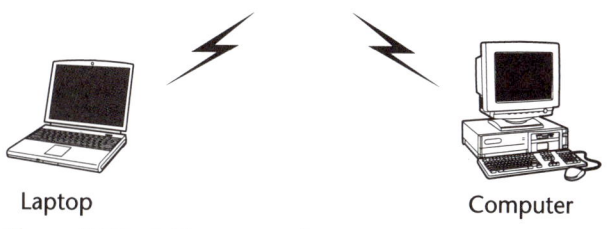

Figure 2.11 Ad-hoc network.

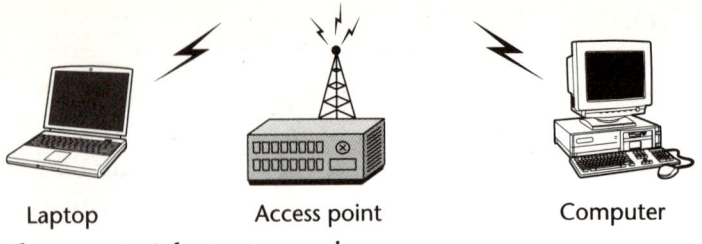

Figure 2.12 Infrastructure mode.

Most wireless LANs are constructed to operate in the infrastructure mode configuration. In bigger networks, an infrastructure mode can be further extended to form distributions systems.

Distribution Service Systems (DSSs)

Distribution systems let wireless LANs be connected to the wired world. A distribution system allows the APs to engage in a hierarchical network configuration, which makes those computers in wireless LANs part of the total network. A distribution system may be created from existing or new technologies. A point-to-point bridge—a network device that interconnects LANs of various types—connecting LANs in two separate buildings could become a DS.

In order for distribution systems to work, they must provide services to the lower-level wireless networks. These services are divided into two sections: distribution system services (DSSs) and station services (SSs).

DSSs provide five basic services: association, reassociation, disassociation, distribution, and integration. The first three services deal with station mobility. If a station is moving within its own BSS or is not moving, the station's mobility is termed no-transition. If a station moves between BSSs within the same ESS, its mobility is termed BSS-transition. If the station moves between BSSs of differing ESSs, it is ESS-transition. A station must affiliate itself with the BSS infrastructure if it wants to use the LAN. The station can do this by associating itself with an AP. Associations are dynamic in nature because stations move, turn on, or turn off. A station can only be associated with one AP. This ensures that the DS always knows where the station is. Association supports no-transition mobility but is not enough to support BSS-transition. The enter reassociation service allows the station to switch its association from one AP to another. Both association and reassociation are initiated by the station, which wants to join the network. Disassociation is when the association between the station and the AP is terminated. Either party can initiate disassociation. A disassociated station cannot send or receive data. That is because it is not supported. A station can move to a new ESS, but it will have to reinitiate connections. Distribution and integration are the remaining DSSs. Distribution is simply getting the data from the sender to the intended receiver. The message is sent to the local AP (input AP) and then distributed through the DS to the AP (output AP) that the recipient is associated with. If the sender and receiver are in the same BSS, the input and output APs are the same. So the distribution service is logically invoked, whether the data is going through the DS or not. Integration is when the output AP is a portal.

Station services are authentication, deauthentication, privacy, and MAC service data unit (MSDU) delivery. With a wireless system, the medium is not exactly bounded as with a wired system. In order to control access to the network, stations must first establish their identity. This is much like trying to enter into a secured facility where you must identify yourself before you are allowed to get inside. In computer networks, before you are allowed a connection, you must first pass a series of tests to ensure that you are who you say you are. In wireless LANs, once a station has been authenticated, it may then associate itself. The authentication relationship may be between two stations inside an IBSS in an ad-hoc network, or to the AP of the BSS in an infrastructure network. All stations start with unauthorized status until they are authenticated. Deauthentication is when either the station or AP wishes to terminate a station's authentication. When this happens, the station is automatically disassociated and the connection-related information on the AP is discarded. Privacy in wireless LANs is achieved through the use of encryption technology. Without encryption, the data is transmitted in the cleartext or plaintext. Data transmitted in cleartext is vulnerable to eavesdropping and tampering by adversaries. In most wireless LANs, privacy is an optional feature and can be enabled, if higher security is desired. To use a wireless LAN in privacy-enabled mode, the stations and AP must be configured to use the same encryption parameters (technology, encryption keys, and so on); otherwise they will not be able to interpret the received data. MSDU delivery ensures that the information in the MAC service data unit is delivered between the media access control service APs. The bottom line is that this authentication is basically a network-wide password. Privacy is ensured if encryption is used.

Existing Wireless LAN Standards

With a wide variety of devices available today, each produced by a different company, manufacturers realized the need to make their devices interoperable with one another or at least to follow a given standard. At first, some vendors introduced wireless LAN solutions based on the proprietary technology; these solutions were not interoperable with devices from other vendors and required entire infrastructure to be purchased from one specific vendor. The IEEE recognized a need for a standard that utilizes the limited wireless RF bandwidth in the most efficient manner.

IEEE 802.11

To address the need for some uniformity in operability of different types of wireless LANs, the IEEE committee responsible for Local Area Network standards and Metropolitan Area Network standards, known as the 802 LAN/MAN Standards Committee, formed a new working group called 802.11 to explore standards for the wireless LANs. In 1997, IEEE drafted the 802.11 standard for wireless local area networking. The IEEE 802.11 standard defines the transmission infrared light and two types of radio transmission within the unlicensed 2.4-GHz frequency band. We examine more about 802.11 standards in the next chapter.

IEEE 802.11b

In 1999, the 802.11b standard was drafted and accepted by the networking industry, and products for wireless networking over the 2.4-GHz frequency began being produced. 802.11b uses the ISM band and operates up to 11 Mbps with a fallback to 5.5, 2, and 1 Mbps. 802.11b uses DSSS as its spread spectrum technology. 802.11b also supports Wired Equivalent Privacy (WEP) for confidentiality of data transmitted over the wireless LAN. 802.11b is also known as wireless fidelity (Wi-Fi). Most wireless LAN device manufacturers and the Wireless Ethernet Compatibility Alliance (WECA) are promoting this standard.

IEEE 802.11a

802.11a is the upcoming product of the IEEE 802.11 working group. The standard was formalized to develop a physical layer that operates in the newly allocated UNII band. This is an extension to 802.11 that applies to wireless LANs and provides up to 54 Mbps in the 5-GHz band. 802.11a uses an orthogonal frequency division multiplexing encoding scheme rather than FHSS or DSSS. Almost all major vendors have now introduced their line of 802.11a devices. Most 802.11a devices are targeted toward the enterprise market.

HomeRF

HomeRF also operates in the same 2.4-GHz ISM band as 802.11b and 2.4-GHz cordless telephones. HomeRF uses FHSS as its spread spectrum technology. HomeRF networks provide a range of up to 150 feet, sufficient to cover the typical home, garage, and yard.

Bluetooth

Bluetooth is one of the most recent wireless standards. Bluetooth is a strong candidate for the personal area network or PAN devices. PAN is defined as a wireless network ranging from a few inches to up to 10 feet; essentially a network around one's personal space. Bluetooth also operates in the ISM band. Current applications for Bluetooth include data synchronization for handheld personal digital assistants, wireless headsets, and similar gadgets.

Are Wireless LANs Risks to Health?

Wireless LAN equipment radiates electromagnetic energy. The health of a living being may be adversely affected by such waves. A good device would provide the lowest possible hazard. Before purchasing or using any device that uses electromagnetic energy, carefully read the equipment manual and look for information regarding the radiated output power of the device. If a device comes with an FCC ID, you can obtain information regarding emission disclosure and frequency usage from the FCC Web site at www.fcc.gov/oet/fccid. At the site, just enter the FCC ID of the device, which consists of the three-character grantee code and the equipment product code, or EPC (up to fourteen characters long) for example:

```
FCC ID: ABC12345678901234
```

Security Risks

Wireless LANs normally use the Wired Equivalent Privacy (WEP) for providing confidentiality of the data transmitted over the air. WEP is a security protocol, specified in the IEEE Wi-Fi standard that is designed to provide a wireless LAN with a level of security and privacy comparable to what is usually expected of a wired LAN. A wired LAN is generally protected by physical security mechanisms (controlled access to a building, for example) that are effective for a controlled physical environment, but may be ineffective for wireless LANs because radio waves are not necessarily bound by the walls containing the network. WEP seeks to establish similar protection to that offered by the wired network's physical security measures by encrypting data transmitted over the wireless LAN. This way even if someone eavesdrops at the wireless packets, he or she will not be successful in understanding the content of the data being transmitted over the wireless LAN.

However, a research group from the University of California at Berkeley recently published a report citing "major security flaws" in WEP that left wireless LANs using the protocol vulnerable to attacks. But the Wireless Ethernet Compatibility Alliance (WECA), an organization formed by major 802.11 equipment manufacturers to promote the use of wireless LANs and perform equipment interoperability among its members, claims that WEP was never intended to be the sole security mechanism for wireless LANs. We cover the security of wireless LANs in much more depth in later chapters.

Summary

In this chapter we briefly described the history of wireless networks. We saw that wireless networks have been in use since the 1950s. We first examined the basic operation of a simple wireless network where we saw how two computers can be connected with each other to form a simple wireless LAN. Then we examined in detail the architecture of a generic wireless LAN. We saw that most wireless LANs operate in the Industrial, Scientific, and Medical (ISM) band; and that in the United States the FCC mandates that such devices must use a spread spectrum technology. We analyzed the different components of wireless LANs. We explored different configurations in which wireless LANs can be used. In the end, we talked about existing standards and saw that 802.11 is perhaps the most appropriate existing standard for wireless LANs. In the next chapter, we examine the IEEE 802.11 standard and its extensions in detail.

CHAPTER 3

The Institute of Electrical and Electronics Engineers (IEEE) 802.11 Standards

The Institute of Electrical and Electronics Engineers (IEEE) 802.11 is a working group of the IEEE 802 LAN/MAN Standards Committee (IEEE 802 LMSC). The goal of the 802.11 Working Group is to develop the physical (PHY) and the media access control (MAC) layer standards for wireless LAN.

In this chapter we examine the wireless standards that the IEEE 802 LMSC has approved and those that are up and coming. Our focus is 802.11, the wireless LAN working group. We explain the major differences between various 802.11 standards, their operation, interoperability, and deployment constraints.

In the paragraphs that follow we discuss a brief history of the IEEE; IEEE working groups responsible for development of wireless LAN standards; a basic overview of 802.11 standard, extensions, and its shortcomings; and a brief comparison of the IEEE 802 wireless standards.

First, to understand the significance of the IEEE and the importance of its involvement in the development of the wireless LAN standards, let's look at the history of the IEEE.

History of IEEE

The existence of the IEEE dates to May 13, 1884, when the American Institute of Electrical Engineers (AIEE) was formed in New York City. AIEE played an active role in the development of electrical industry standards, which focus primarily on the wired communications, light, and power systems. In the early 1900s the Society of Wireless and Telegraph Engineers and the Wireless Institute, two separate organizations working on wireless communication standards, merged to form the Institute of Radio Engineers (IRE). Though the majority of work done by the IRE was radio communications related, it heavily utilized the advancement in electronics and electrical engineering—an area that was the primary focus of the AIEE. The work done by both the AIEE and the IRE was similar in many respects; hence, many members of the IRE were also members of the AIEE. Recognizing the common goals that both organizations had, their leaders decided to merge the two to form one organization, which would perform the tasks performed by both organizations. The two organizations finally merged on January 1, 1961, to form IEEE. Since 1961, IEEE has played an extremely important role in electrical industry standards development and academics. Today, IEEE has over 377,342 members around the world, its standards are widely accepted, and it publishes over 75 journals and magazines that define the future of the electrical industry.

Within IEEE, most standards-related work is performed by its committees. These committees normally have working groups that deal with a committee-assigned subarea. Depending on the complexity, working groups often designate task groups that do most of the groundwork. The working group first approves the work of the task group, which finally becomes a standard pending approval from the government agencies (if necessary) and the committee that work group reports to. Today, almost all computer network standards are IEEE-compliant.

The IEEE 802 LAN/MAN Standards Committee (IEEE 802 LMSC) was formed in 1980 to develop and propose standards for LANs. The most commonly used LAN standards 802.3 (Ethernet or CSMA/CD) and 802.5 (token ring) were both developed by IEEE. Today, there are 17 different working groups that operate under the authority of IEEE 802 LMSC. Each working group is named after its standards committee and is identified by a numerical value. For example, 802.11 is an IEEE 802 LMSC working group for wireless LAN.

IEEE 802 Wireless Standards

The scope of the IEEE 802 LMSC Committee has grown since its inception in 1980. Today, there are three basic wireless working groups within the IEEE 802

LMSC: the IEEE 802.11 for wireless LANs, the IEEE 802.15 for personal area networks (PANs), and the IEEE 802.16 for broadband wireless solutions.

The 802.11 Working Group

The IEEE 802.11 was formed in July 1990 to develop CSMA/CA, a variation of CSMA/CD (Ethernet)-based wireless LANs. The working group produced the first 802.11 standard in 1997, which specifies wireless LAN devices capable of operating up to 2 Mbps using the unlicensed 2.4-GHz band. Currently, the working group has nine basic task groups and each is identified by a letter from a to i. Following are the current 802.11 task groups and their primary responsibilities:

802.11a. Provides a 5-GHz band standard for 54-Mbps transmission rate.

802.11b. Specifies a 2.4-GHz band standard for up to 11-Mbps transmission rate.

802.11c. Gives the required 802.11-specific information to the ISO/IEC 10038 (IEEE 802.1D) standard.

802.11d. Adds the requirements and definitions necessary to allow 802.11 wireless LAN equipment to operate in markets not served by the current 802.11 standard.

802.11e. Expands support for LAN applications with Quality of Service requirements.

802.11f. Specifies the necessary information that needs to be exchanged between access points to support the P802.11 DS functions.

802.11g. Develops a new PHY extension to enhance the performance and the possible applications of the 802.11b compatible networks by increasing the data rate achievable by such devices.

802.11h. Enhances the current 802.11 MAC and 802.11a PHY with network management and control extensions for spectrum and transmit power management in 5-GHz license exempt bands.

802.11i. Enhances the current 802.11 MAC to provide improvements in security.

We will discuss the 802.11 working group family of standards in much detail in the section *The 802.11 Family of Standards* later in this chapter.

The 802.15 Working Group

The IEEE 802.15 Working Group first met in July 1999. The working group develops standards and recommends practices for short-distance wireless networks known as wireless personal area networks (WPANs). These WPANs address the needs of personal digital assistants (PDAs), personal computers (PCs), cell phones, and wireless payment systems. The WPAN-compliant devices are supposed to operate within the personal operating space (POS) that typically extends about a radius of 5 meters from a WPAN device. A number—for example, 802.15.1—denotes the projects and the task groups of 802.15. The working group currently has the following four projects:

802.15.1. A WPAN standard for Bluetooth

802.15.2. A coexistence guideline for license-exempt devices

802.15.3. A high-rate WPAN standard

802.15.4. A low-rate WPAN standard

The most widely implemented standard of the 802.15 Working Group is 802.15.1, which uses Bluetooth technology and operates in the 2.4-GHz ISM band.

The 802.16 Working Group

The 802.16 Working Group was formed in July of 1999 for developing standards and recommending practices for the development and deployment of fixed broadband wireless access systems. The working group has the following three projects:

802.16. Air Interface for 10-66 GHz Recommended practice for coexistence among 802.16 and 802.16a devices

802.16a. Amendments to the MAC layer and an additional PHY layer for 2-11 GHz licensed frequencies

802.16b. Amendments to the MAC layer and an additional PHY layer, license-exempt frequencies, with a focus on 5-6 GHz

The 802.11 Family of Standards

802.11 refers to a family of specifications developed by the IEEE for wireless LAN technology. The original 802.11 standard specifies an over-the-air interface between a wireless client and a base station or between two wireless clients. The IEEE accepted the specification for 802.11 in 1997. The task groups within the 802.11 working group have produced few extensions to the original specification. The products of these extensions are named after the task group and the original specification—for example, 802.11b is an extension developed by the task group b. The most popular extensions of 802.11 specifications are 802.11b, 802.11a, and 802.11g.

In this section, we first look at the 802.11 standard, and then we examine the popular extensions in detail.

The 802.11 Standard Details

The 802.11 standard specifies wireless LANs that provide up to 2 Mbps of transmission speed and operate in the 2.4-GHz Industrial, Scientific, and Medical (ISM) band using either frequency-hopping spread spectrum (FHSS) or direct-sequence spread spectrum (DSSS). The IEEE approved this standard in 1997. The standard defines a physical layer (PHY), a medium access control (MAC) layer, the security primitives, and the basic operation modes.

The Physical Layer

The 802.11 standard supports both radio frequency- and infrared-based physical network interfaces. However, most implementations of 802.11 use radio frequency, and we only discuss the radio frequency-based physical interface here.

802.11 Frequency Bandwidth

802.11 standard-compliant devices operate in the unlicensed 2.4-GHz ISM band. Due to the limited bandwidth available when the electromagnetic spectrum is used for data transmission, many factors have to be considered for reliable, safe, and high-performance operation. These factors include the technologies used to propagate signals within the RF band, the time that a single device is allowed to have an exclusive transmission right, and the modulation scheme. For these reasons, FCC regulations require that radio frequency systems must use spread spectrum technology when operating in the unlicensed bands.

Spread Spectrum Technology

The 802.11 standard mandates using either DSSS or FHSS. In FHSS, the radio signal hops within the transmission band. Because the signal does not stay in one place on the band, FHSS can elude and resist radio interference. DSSS avoids interference by configuring the spreading function in the receiver to concentrate the desired signal, and to spread out and dilute any interfering signal.

Direct-Sequence Spread Spectrum (DSSS)

In DSSS the transmission signal is spread over an allowed band. The data is transmitted by first modulating a binary string called spreading code. A random binary string is used to modulate the transmitted signal. This random string is called the spreading code. The data bits are mapped to a pattern of "chips" and mapped back into a bit at the destination. The number of chips that represent a bit is the spreading ratio. The higher the spreading ratio, the more the signal is resistant to interference. The lower the spreading ratio, the more bandwidth is available to the user. The FCC mandates that the spreading ratio must be more than 10. Most products have a spreading ratio of less than 20. The transmitter and the receiver must be synchronized with the same spreading code. Recovery is faster in DSSS systems because of the ability to spread the signal over a wider band.

Frequency-Hopping Spread Spectrum (FHSS)

This spread spectrum technique divides the band into smaller subchannels of usually 1 MHz. The transmitter then hops between the subchannels sending out short bursts of data for a given time. The maximum amount of time that a transmitter spends in a subchannel is called the dwell time. In order for FHSS to work correctly, both communicating ends must be synchronized (that is, both sides must use the same hopping pattern), otherwise they lose the data. FHSS is more resistant to interference because of its hopping nature. The FCC mandates that the band must be split into at least 75 subchannels and that no subchannel is occupied for more than 400 milliseconds. Debate is always ongoing about the security that this hopping feature provides. Since there are only 75 subchannels available, the hopping pattern has to be repeated once all the 75 subchannels have been hopped. HomeRF FHSS implementations select the initial hopping sequence in a pseudorandom fashion from among a list of 75 channels without replacement. After the initial 75 hops, the entire sequence is repeated without any replacement or change in the hopping order. An intruder could possibly compromise the system by monitoring and recording the hopping sequence and then waiting till the whole sequence is repeated. Once the hacker confirms the hopping pattern, he or she can predict the next subchannel that hopping pattern will be using thereby defeating the hopping advantage altogether. HomeRF radios, for example, hop through each of the 75 hopping channels at a rate of 50 hops per second in a total of 1.5 seconds, repeating the same pattern each time, enabling a hacker to guess the hopping sequence in 3 seconds. Nevertheless, this technique still provides a high level of security in that expensive equipment is needed to break it. Many FHSS LANs can be colocated if an orthogonal hopping sequence is used. Since the subchannels in FHSS are smaller than DSSS, the number of colocated LANs can be greater with FHSS systems. The most commonly used standard based on FHSS is HomeRF.

The MAC Layer

The MAC layer controls how data is to be distributed over the physical medium. The main job of the MAC protocol is to regulate the usage of the medium, and this is done through a channel access mechanism. A channel access mechanism is a way to divide the available bandwidth resource between subchannels—the radio channel—by regulating the use of it. It tells each subchannel when it can transmit and when it is expected to receive data. The channel access mechanism is the core of the MAC protocol. With most wired LAN using the Carrier Sense Multiple Access with Collision Detection (CSMA/CD) it was a logical choice for the 802.11 Working Group to apply the CSMA/CD technology when developing the MAC layer for the 802.11 standard.

The working group chose the Carrier Sense Multiple Access with Collision Avoidance (CSMA/CA), a derivative of CSMA/CD, as the MAC protocol for the 802.11 standard. CSMA/CA works as follows: The station listens before it sends. If someone is already transmitting, it waits for a random period and tries again. If no one is transmitting, then it sends a short message. This message is called the ready-to-send message (RTS). This message contains the destination address and the duration of the transmission. Other stations now know that they must wait that long before they can transmit. The destination then sends a short message, which is the clear-to-send message (CTS). This message tells the source that it can send without fear of collisions. Upon successful reception of a packet, the receiving end transmits an acknowledgment packet (ACK). Each packet is acknowledged. If an acknowledgment is not received, the MAC layer retransmits the data. This entire sequence is called the four-way handshake.

802.11 Security

IEEE 802.11 provides two types of data security authentication and privacy. Authentication is the means by which one station verifies the identity of another station in a given coverage area. In the infrastructure mode, authentication is established between an AP and each station. When providing privacy, a wireless LAN system guarantees that data is encrypted when traveling over the media.

There are two types of authentication mechanisms in 802.11: open system or shared key. In an open system, any station may request authentication. The station receiving the request may grant authentication to any request, or to only those from stations on a preconfigured user-defined list. In a shared-key system, only stations that possess a secret encrypted key can be authenticated. Shared-key authentication is available only to systems having the optional encryption capability.

The 802.11 standard mandates the use of Wired Equivalent Privacy (WEP) for providing confidentiality of the data transmitted over the air at a level of security comparable to that of a wired LAN. WEP is a security protocol, specified in the IEEE wireless fidelity (Wi-Fi) standard that is designed to provide a wireless LAN with a level of security and privacy comparable to what is usually expected of a wired LAN. WEP uses the RC4 Pseudo Random Number Generator (PRNG) algorithm from RSA Security, Inc. to perform all encryption functions. A wired LAN is generally protected by physical security mechanisms (for example, controlled access to a building) that are effective for a controlled physical environment, but they may be ineffective for wireless LANs because radio waves are not necessarily bounced by the walls containing the network. WEP seeks to establish protection similar to that offered by the wired network's physical security measures by encrypting data transmitted over the wireless LAN. This way even if someone listens in to the wireless packets, that eavesdropper will not be successful in understanding the content of the data being transmitted over the wireless LAN.

Operating Modes

The 802.11 standard defines two operating modes: the ad hoc and the infrastructure mode. To understand how an 802.11 wireless LAN operates, let's understand the basic terminologies used to describe the two modes.

Terminologies

The terminologies describing the two operating modes include a station, an independent basic service set (IBSS), a basic service set (BSS), an extended service set (ESS), an access point (AP), and a distribution system (DS). Each of these is discussed in the paragraphs that follow.

An 802.11 Station

An 802.11 station is defined as an 802.11-compliant device. This could be a computer equipped with an 802.11-compliant network card.

Basic Service Set (BSS)

A BSS consists of two or more stations that communicate with each other.

An Access Point (AP)

An AP is a station in an 802.11 wireless LAN that routes the traffic between the stations or among stations within a BSS. The AP can simply be a routing device with 802.11 capabilities. An AP must have a network address, it must act like a regular station on the network, and it must be addressable by the other stations on the network. An AP periodically sends beacon frames to announce its

The Institute of Electrical and Electronics Engineers 802.11 Standards

presence, it provides new information to all stations, authenticates users, manages transmitted data privacy, and keeps stations synchronized with the network.

Independent Basic Service Set (IBSS)

A BSS that stands alone and is not connected to an AP is called an independent basic service set (IBSS).

Distribution System (DS)

A distribution system interconnects multiple APs, forming a single network. A distribution system, therefore, extends a wireless network. The 802.11 standard does not specify the architecture of a DS, but it does require that a DS must be supported by 802.11-compliant devices.

Now that we know the basic terminologies, let's look at the operating modes of an 802.11 wireless LAN.

802.11 Ad-Hoc Mode

When a BSS-based network (two or more stations connected with each other over wireless) stands alone and is not connected to an AP, it is known as an ad-hoc network. An ESS is formed when two or more BSSs operate within the same network. An ad-hoc network is a network where stations communicate only peer-to-peer. An example of a wireless LAN operating in ad-hoc mode would be a LAN with two computers communicating with each other using a wireless link.

Infrastructure Mode

An 802.11 network is known to be operating in infrastructure mode when two or more BSSs are interconnected using an access point. Access points act like hubs for wireless stations. An access point routes the traffic between the two BSSs. An access point is sometimes connected to a wired network to provide wired network resources to the wireless stations. Each BSS becomes a component of an extended, larger network. An access point is a station, thus addressable. So data moves between the BSS and the wired network with the help of these access points. A wireless LAN consisting of two computers and an AP, with each computer equipped with wireless LAN adapters, is an example of a wireless LAN operating in the infrastructure mode.

Roaming

The 802.11 standard does not define a standard mechanism for roaming. Roaming is a feature of wireless LAN that enables a station to travel between the APs without any gap or loss of connectivity during transit. Though 802.11 does not define how roaming should be performed, it does provide the basic support functions that can be used to perform roaming. It is up to the individual implementers to choose how to support roaming in their devices. In most cases, the station association and disassociation services are used to enable the roaming feature. The APs are installed such that they barely overlap their operating space. When a roaming user approaches the functional boundary of the AP it is currently associated with, the network adapter, upon realization of weaker signal, starts looking for other APs in the area. If the network adapter finds a stronger signal in the newly discovered APs, it disassociates itself from the AP with which it was associated and associates itself with the newly discovered AP.

The 802.11 Extensions

The 802.11 Working Group realized that the initial standard that was passed in 1997 would not be sufficient to attract implementers. Therefore, the working group established various task groups with the responsibilities to develop different extensions to the 802.11 standard. The idea behind having different task groups is to develop standards for different types of usage scenarios that still conform to a basic set of operating rules and are still interoperable to a certain extent. The most promising standards at this time include 802.11b, 802.11a, 802.11g, and 802.11e. We discuss the extensions in the order of their popularity, development status, and general acceptance.

802.11b

802.11b is an extension to 802.11 that operates at speeds up to 11 Mbps transmission (with a fallback to 5.5, 2, and 1 Mbps) in the 2.4-GHz band and uses only DSSS. 802.11b is also known as 802.11 high rate or wireless fidelity (Wi-Fi).

Enhancements Offered by 802.11b over 802.11

The 802.11b extension was the product of the 802.11 task group b and was approved in 1999. 802.11b ratifies to the original 802.11 standard, allowing wireless functionality comparable to Ethernet. The 802.11b standard operates up to 11 Mbps, whereas the base 802.11 standard supported speeds of up to 2 Mbps.

802.11b Applications

802.11b is the most widely deployed wireless LAN standard. 802.11b is currently available in the market. Now with operating speeds up to 11 Mbps, it is far more practical to use the wireless LANs than the conventional wired LANs. It is being used in Small Office Home Office (SoHo) environments, enterprises, and by Wireless Internet Service Providers (WISPs).

Small Office Home Office (SoHo)

802.11b is very attractive to home users and to those who operate a small business from home. Users enjoy the instant networking that was very impractical in the recent past. Now, no cumbersome wiring or understanding of the cable types is needed. Just buy one or more 802.11b-compliant network cards and an AP. Install according to the manufacturer's instructions and you have a functional computer network. This ease of deployment is making 802.11-based wireless LANs a popular alternative to the wired LANs for SoHo environments. With 802.11b-compliant APs that come with built-in broadband support, sharing an Internet connection among multiple users is now easier than before. Most APs these days come with DSL or cable modem connectivity that provides the ability to connect a wireless LAN to the Internet.

Enterprise

Enterprise users can be more mobile with a wireless LAN that is constructed using 802.11b networking devices. These networks provide scalability and enable users to move about within the organization without worrying about the wiring and other physical constraints.

Wireless Internet Service Providers (WISPs) and Community Networks

Internet Service Providers (ISPs) are seeing a great business opportunity in providing wireless Internet access services to mobile users. Today, many Internet cafes, coffee shops, airports, and parks are equipped with 802.11b APs. These APs are operated by the private WISPs who charge the users for accessing the Internet using their computer. All a user has to bring to such a location is a computing device equipped with an 802.11b network card and a valid credit card to pay for the WISP access fees.

802.11b Limitations

802.11b is haunted by the possibility of interference in the 2.4-GHz frequency band in which it operates. However, the 2.4-GHz frequency is already crowded and will soon be more so. Microwave ovens operate at 2.4 GHz and can deter the performance of 802.11 wireless networks. Many powerful cordless phones also operate at the 2.4-GHz frequency. If you use 802.11b networking products, forget about using these phones in the same area.

An even greater threat to 802.11b stability is just around the corner. Bluetooth, the short-range wireless networking standard, which also operates in the 2.4-GHz range, is slated to coexist with wireless LANs. Bluetooth is not bothered a bit by 802.11b signals, but not vice versa. Depending on the proximity and number of devices, Bluetooth can have a negative impact on the performance of an 802.11b connection due to electromagnetic interference caused by the Bluetooth devices. Fortunately, Bluetooth-enabled devices are used for transmission of a small amount of data—for example, synchronization of a phonebook in a cell phone with a desktop computer—over short periods of time and generally do not cause major network problems. Most interference can be avoided by configuring the 802.11b equipment to choose channels that operate on one end of the spectrum and Bluetooth devices to operate on the other. That said, however, a visitor equipped with Bluetooth equipment configured to operate in overlapping frequency can still cause limited interference.

802.11b Interoperability and Compatibility with 802.11

802.11b devices are backward compatible with 802.11 implementations, which use the DSSS as their spectrum technology. Therefore, 802.11b devices operate at lower speeds when they are connected to an 802.11 network. 802.11b devices are not compatible with the HomeRF devices because HomeRF uses the FHSS standard.

802.11a

The 802.11a standard was approved in December 1999, right around the same time as 802.11b was approved. 802.11a is an extension to 802.11, which operates at speeds of up to 54-Mbps transmission rate (with a fallback to 48, 36, 24, 18, 12, and 6 Mbps) in the more recently allocated 5-GHz Unlicensed National Information Infrastructure (UNII) band. 802.11a uses an Orthogonal Frequency Division Multiplexing (OFDM) encoding scheme as its spread spectrum technology.

802.11a is to 802.11b networking what 100 Mbps was to the 10-Mbps Ethernet. The acceptance of the 802.11a standard lagged behind the 802.11b because of the relative complexity of the standard and the cost of equipment that it incurs. In addition, 802.11a networks are incompatible with the 802.11b networks due to the difference in the radio frequency band used by 802.11a (802.11b uses 2.4 GHz whereas 802.11a uses 5 GHz), and the speeds they operate at (802.11b has a maximum operating speed of 11 Mbps whereas 802.11a operates at up to 54 Mbps).

Enhancements Offered by 802.11a over 802.11

In the United States, 802.11a operates in three unlicensed radio frequencies in the 5-GHz radio band, instead of the 2.4-GHz frequency used by 802.11. At the 2.4-GHz frequency, only three channels can be used simultaneously; 802.11a supports eight simultaneous channels, and full bandwidth is available within each channel. The additional channels mean that more users can share the same frequency.

802.11a and Orthogonal Frequency Division Multiplexing (OFDM)

OFDM technique distributes the data to be transmitted into smaller pieces, which are simultaneously transmitted over multiple frequency channels that are spaced apart. This spacing provides the orthogonality that prevents the demodulators from seeing frequencies other than their own.

When transmitting data using OFDM, the data is first divided into frames and a mathematical algorithm known as Fast Fourier Transformation (FFT) is applied to the frame, then OFDM parameters (for example, timing) are added. An Inverse Fast Fourier Transformation (IFFT) is then applied on each frame. The resulting frames are then transmitted over the designated frequencies. A receiver performs the inverse operations to get the transmitted data by performing FFT on the frames.

The benefits of OFDM are high spectral efficiency, resiliency to RF interference, and lower multipath distortion.

802.11a Applications

Currently, not many devices are available in the markets that comply with the 802.11a standard. With growing usage of 802.11b, 802.11a is slow to gain the market share that it deserves because implementation choices and vendor support were limited until this year. Still, 802.11a is gaining acceptance in the enterprise market. Several large equipment vendors have announced 802.11a implementations. 802.11a is being compared with 802.11a like fast Ethernet is compared with Ethernet. Because 802.11a operates in 5 GHz, it can coexist with 802.11b networks without causing any interference. 802.11a is being used to connect network backbones in small enterprise environments and the applications that require high bandwidth.

Enterprise users normally desire a higher level of reliability and speed than SoHo or home users do. 802.11a is well suited for such scenarios. 802.11a operates at speeds up to 54 Mbps and is less vulnerable to the interference caused by devices competing for the bandwidth in the 2.4-GHz band.

802.11a Interoperability and Compatibility with 802.11

The 802.11a-compliant devices are not directly compatible with the original 802.11 standard or the 802.11b extension. The primary reason is the RF band in which 802.11a operates. The original 802.11 specification calls for devices that would operate in the 2.4-GHz ISM band, whereas 802.11a devices operate in the 5-GHz UNII band. This gives the 802.11a devices the freedom of operating in an RF band with a smaller number of devices. In addition, 802.11a devices use OFDM as their spread spectrum technology versus FHSS or DSSS, which 802.11 originally mandated. However, 802.11a uses the same MAC layer (CSMA/CA) as the original 802.11 specification recommended. The usage of the same MAC-level protocol makes 802.11a devices interoperable at the MAC layer with other 802.11 devices.

802.11g

IEEE 802.11 LMSC adopted the 802.11g standard in late 2001. The 802.11g standard is still under development. The 802.11g standard operates in the 2.4-GHz band and provides speeds up to 54 Mbps (with a fallback to 48, 36, 24, 18, 11, 5.5, 2, and 1 Mbps). The 802.11g differs from 802.11b because it can optionally use OFDM (802.11g draft mandates that OFDM be used for speeds above 20 Mbps).

Enhancements Offered by 802.11g over 802.11

The most important enhancement offered by 802.11g is its higher speed. The ability to operate up to 54 Mbps provides 802.11g a higher edge over other 802.11 compliant devices that operate in the 2.4-GHz band. The support of OFDM is another enhancement that 802.11g maintains over the basic 802.11 standard. OFDM will allow 802.11g to operate in a more efficient manner than the rest of the 802.11-compliant 2.4-GHz devices.

802.11g Applications

The 802.11g devices are not available yet. However, electrical industry analysts predict that when 802.11g becomes available, it would be the only choice that users would consider, as it provides the direct upgrade path and interoperability with the 802.11b standard devices.

SoHo

SoHo users would prefer purchasing 802.11g devices to the currently available 802.11b. Again all credit goes to the backward compatibility and the higher speed that 802.11g provides. However, those users who are not very computer savvy might still go with the 802.11b solutions because they might be cheaper.

Enterprise

Enterprise users would be the primary targets for the 802.11g-compliant devices. The devices built on the 802.11g standard would be a logical upgrade path for the current 802.11b users. The backward compatibility of 802.11g allows 802.11b devices to coexist in the same network environment. This will enable an enterprise IT to selectively upgrade the 802.11b users to 802.11g.

WISPs

WISPs would find it attractive to deploy the 802.11g devices, as this would enable a broader user base to access the services they offer.

802.11g Interoperability and Compatibility with 802.11

Since 802.11g is backward compatible with the 802.11b standard, industry critics are looking forward to its arrival in the marketplace. The 802.11g devices would be the logical choice for the current users of 802.11b who are seeking higher speeds and are willing to upgrade only to a standard that is backward compatible. The 802.11g standard satisfies such users by operating in the 2.4-GHz band and supporting DSSS for speeds up to 20 Mbps (to be compatible with 802.11b, 802.11g needs to operate in DSSS in only up to 11 Mbps).

The 802.11g devices would directly compete with the 802.11a devices, as 802.11g provides the backward compatibility that 802.11a does not. However, 802.11a operates in a less congested RF band than 802.11g does.

802.11g Limitations

Though 802.11g devices would provide higher speed than the currently available 802.11b devices, it still suffers the interference issue with other devices operating in the same RF band, primarily the Bluetooth devices.

802.11 Shortcomings

Security is perhaps the biggest shortcoming of the 802.11 standard. Several papers have been written on the weaknesses of the WEP-based security that 802.11 provides. A research group from the University of California at Berkeley recently published a report citing "major security flaws" in WEP that left wireless LANs using the protocol vulnerable to attacks [2]. But Wireless Ethernet Compatibility Alliance (WECA) claims that WEP was never intended to be the sole security mechanism for wireless LANs [3].

The 802.11 standard does not define any direct support for load balancing. This reduces the scalability of the 802.11 systems. Without load balancing, a given region can operate with only one AP and all users in the region must share the bandwidth of a single AP.

Wireless Standards Comparison

Currently, there are four 802 wireless standards that are gaining popularity: 802.11b, 802.11a, 802.11g, and 802.15.1. Other popular existing standards include HomeRF and Bluetooth. Table 3.1 shows their basic properties.

Table 3.1 Popular 802 Wireless Standards

STANDARD	RF BAND	MAXIMUM SPEED
802.11b	2.4-GHz ISM Band	11 Mbps
802.11a	5-GHz UNII Band	54 Mbps
802.11g	2.4-GHz ISM Band	54 Mbps
802.15.1	2.4-GHz ISM Band	Approximately 700 Kbps
HomeRF	2.4-GHz ISM Band	10 Mbps
Bluetooth	2.4-GHz ISM Band	Approximately 700 Kbps to 1 Mbps

Summary

The 802.11 working group has produced two widely accepted standards: 802.11b and 802.11a. The 802.11g standard is new and is still in the approval process. The 802.11b standard is most popular and operates at speeds of up to 11 Mbps in the 2.4-GHz ISM band. The 802.11a can operate up to 54 Mbps in the 5-GHz UNII band. The 802.11g will operate in the 2.4-GHz ISM band with speeds up to 54 MHz. All 802.11 standards follow the same MAC-layer protocol. This makes the 802.11 devices MAC-layer compatible.

802.11g will provide the upgrade path to most current 802.11b users. Installations that desire high performance would prefer the 802.11a extension because it operates in a relatively newer unlicensed band with few devices, hence with fewest troubles when it comes to interference.

In Chapter 4, "Is Wireless LAN Right for You?" we discuss the benefits of using wireless LANs for various deployment scenarios, costs associated with wireless LANs, deployment issues, and general health concerns. We hope that the next chapter helps you to decide whether or not wireless LANs are right for you.

CHAPTER 4

Is Wireless LAN Right for You?

With the growing use of computers and the popularity of the Internet, it has become viable to deploy LANs in places where we never thought we would need a LAN. Today, LANs are being used in industrial manufacturing, offices, small businesses, and at homes. Wireless networking has taken LAN connectivity a step further. Now, with wireless networking, LANs have become far more flexible than they used to be. Wireless LANs are easier to build than conventional wired LANs and provide mobility to LAN users. Wireless LANs are being used to connect mobile devices, such as personal digital assistants (PDAs) and laptop computers, with stationary computers, such as desktop computers. Wireless networking equipment is also being used to connect separate buildings as well as extending the reach of the Internet and the virtual private networks (VPNs) across several miles in remote areas where wired infrastructure is sparse.

In this chapter we discuss the different aspects of a wireless LAN that directly impact the feasibility for SoHo, enterprise, and WISP deployment scenarios. We talk about the benefits, deployment scenarios, costs associated, deployment issues, bandwidth and network congestion, security, and health concerns of the wireless LANs. If you decide that wireless LANs are not suitable for you, you should look at Chapter 1, "Networking Basics," which describes a basic wired LAN to study whether wired LANs satisfy your needs.

Benefits of Wireless LANs

The primary advantage that wireless LANs have over wired networks is that they do not require wires and can be set up quickly in areas where wiring costs can be prohibitive. The advent of wireless LANs has provided us with a greater level of flexibility on how we configure our computing equipment and environment than the wired LANs. You no longer need separate modems, black-and-white printers, color printers, scanners, CD-ROM readers/writers, and other devices for every computer in your home or office. You also do not need to go through the hassle of keeping multiple copies of files when sharing a document.

When deciding whether a wireless network is right for you, you should first make sure that you do indeed need a LAN. Though LANs provide some very useful services, they incur installation and maintenance costs. To justify your need for a LAN, you should have at least one computer, and one or more of the following should apply to you:

- You want to share files across computers.
- You intend to share a printer among computers.
- Only one Internet connection is available, and you want to share it across two or more computers.
- You intend to share a new type of device that connects to a LAN and make its services available to all the computers on the given LAN—for example, a computer controlled telescope.
- You are willing to spend a decent amount of money to build a network.
- Your workstations and other network devices need to be mobile and not tied down to a particular location.
- Physical limitations prohibit running network cables and drops.
- Lease or other restrictions do not allow for installation of a wiring plant.
- You need to deploy networks in open spaces where you expect a lot of foot traffic, and network wires and equipment would cause additional safety issues.
- You temporarily need a LAN, for example, at a research site.

In today's computing environments, devices, data, and resources are often distributed across multiple points on a network and are accessible from any authorized workstation in that network. Wireless LAN takes these capabilities to the next level by adding mobility to the workstations and network devices.

Within a wireless LAN, the workstations are not limited to a single position in the building but can be moved around while they continue to function. Powerful portable computers and network devices can be carried around a building or campus while they continue to communicate with mission-critical servers and other computers on the rest of the network, sharing information.

Deployment Scenarios

Wireless LANs can be deployed in many different deployment scenarios. Each deployment scenario has a different set of needs. In this section we restrict our focus to small office home office (SoHo), enterprise, and Wireless Internet Service Providers (WISP) scenarios.

Small Office Home Office (SoHo)

Small office home office (SoHo) deployment generally involves either a home LAN, a LAN at a home-based office, or a LAN at a small business. Wireless LANs are rapidly becoming networks of choice for these uses because of their low cost and lack of wiring needs. Setting up wired LANs requires complex wiring generally running to a central point, which is not only costly but in some cases, such as apartments or older homes, almost impossible.

In SoHo environments, the number of computers in a LAN is typically very small. These LANs normally contain between 2 and 10 computers. They are normally used to share files, printers, and data backup devices. Nowadays it is also very common for SoHo networks to share a single Internet connection. Under most circumstances, these networks do not require high security. The speed requirement is nominal, and the budget is small. Therefore, for the SoHo environment, a suitable LAN would be one that is not too complex, has a reasonable level of security, provides the ability to connect with the Internet, and does not require a major investment.

In a SoHo or a home network there may be several computers, a color printer, a black-and-white laser printer, a scanner, several CD-ROM readers, a CD-ROM writer, and a modem (see Figure 4.1). Using a wireless LAN, these resources can be shared efficiently, and you do not need to purchase and install every device for every computer. You can scan a picture from the scanner connected to the desktop in your child's bedroom to the file server (a computer on the LAN with a high-capacity shared hard disk) in your home office that also has the color printer attached it. Then you go to the family room and use the imaging software on your notebook to edit and enhance the picture while you recline in your favorite chair and watch TV surrounded by your loved ones. After completing your first draft, you print the file on the printer attached to the server in your office and review it. You then email the picture to your partner through the Internet-sharing device and cable modem; you also leave a note for your assistant with the file name. When your assistant comes in the next day, he or she opens the file that you saved on the server from his or her workstation and makes the final changes. Over the weekend your friends come over with their laptops and 802.11b Wi-Fi cards and you play network games over the wireless LAN.

Enterprise

Enterprise networks are generally comprised of a larger number of computers, security systems, file-storage and archiving systems, many workstations and laptops, several servers, multiple printers and scanners as well as presentation systems participating in a network. In industrial complexes and manufacturing plants, there may be machinery that needs to communicate with central servers. Enterprise networks are typically divided into several workgroups. The security requirements are very high, the users need to be authenticated, the data and resources have to be protected not only from outsiders but there is also the need to have proper access control for authorized users. The speed and bandwidth requirements are also high, and the network needs to be properly segmented to reduce the network traffic. An enterprise network can also span across multiple floors, multiple buildings, and multiple locations. There may be several Internet and VPN connection lines linking a network with other parts of the enterprise network. There is also the need for covering the complete office area without any dead zones (an area without a network signal) as well as allowing the users to roam freely between floors, in the campus, and across locations.

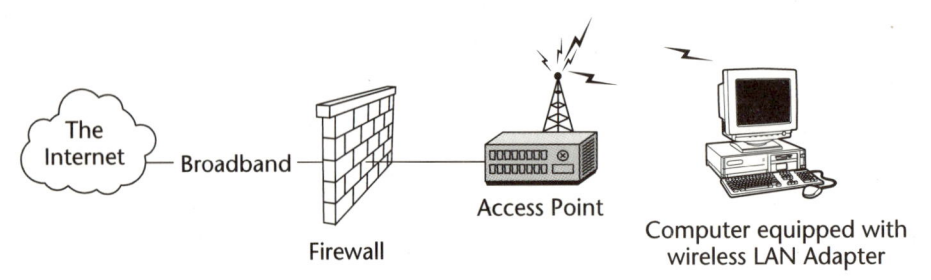

Figure 4.1 A SoHo wireless LAN setup.

Wireless LANs provide the opportunity for enterprises to provide greater mobility to their computer users as well as to lower costs for connecting work areas across buildings and floors (see Figure 4.2). There is no longer a need to run expensive cabling between floors and buildings. This is even more useful in industrial and warehouse situations, where there is an even greater need for mobility for monitoring and data-gathering devices such as inventory scanners. Automobile rental companies have long used wireless networks to check in and check out cars. In offices, wireless networks open the possibility of configuring more flexible workspaces. Many organizations using the wireless LANs provide roaming offices to their employees. In roaming offices, employees do not have fixed offices but use the available space on a per-need basis. In project-oriented workplaces, knowledge workers may need to work in several workgroups during the course of the same day. Using wireless LANs, these workers can get together and collaborate without losing productivity. Knowledge workers no longer need to be tied to their desks to access the data they need. The participants in the meetings can bring their portable computing devices to the meetings. Wireless networking is also changing the structure of meetings. Participants often "chat" in smaller groups and carry out side "conversations" and exchange information privately using their portable devices connected to the network without disturbing the main meeting. There are now 802.11b-based wireless presentation devices coming on the market that allow corporate users to prepare presentations on their workstations and then deliver them without having to deal with the wires on projectors that are permanently attached to wireless receivers. One can expect these receivers to be integrated in the projectors as time passes. We are all familiar with going into a meeting and then waiting for the presenter to connect their computers to the projectors and fiddle with the projectors until they get started.

Wireless Internet Service Providers (WISPs)

Wireless ISPs, or WISPs, are growing very rapidly across the country. Their greatest penetration seems to be in remote or rural areas. As the demand for broadband grows, so grows the gap in availability between urban and remote rural areas. In urban areas, with rapidly declining costs of wireless equipment, opportunities are developing for WISPs to provision buildings without the need for the expensive wiring. There are several WISPs providing services at major hotels, airport terminals, and restaurants. The Wireless ISPs have a higher need for authentication so that only authorized users can access their systems. Generally their security needs are moderate and not as high as enterprise networks.

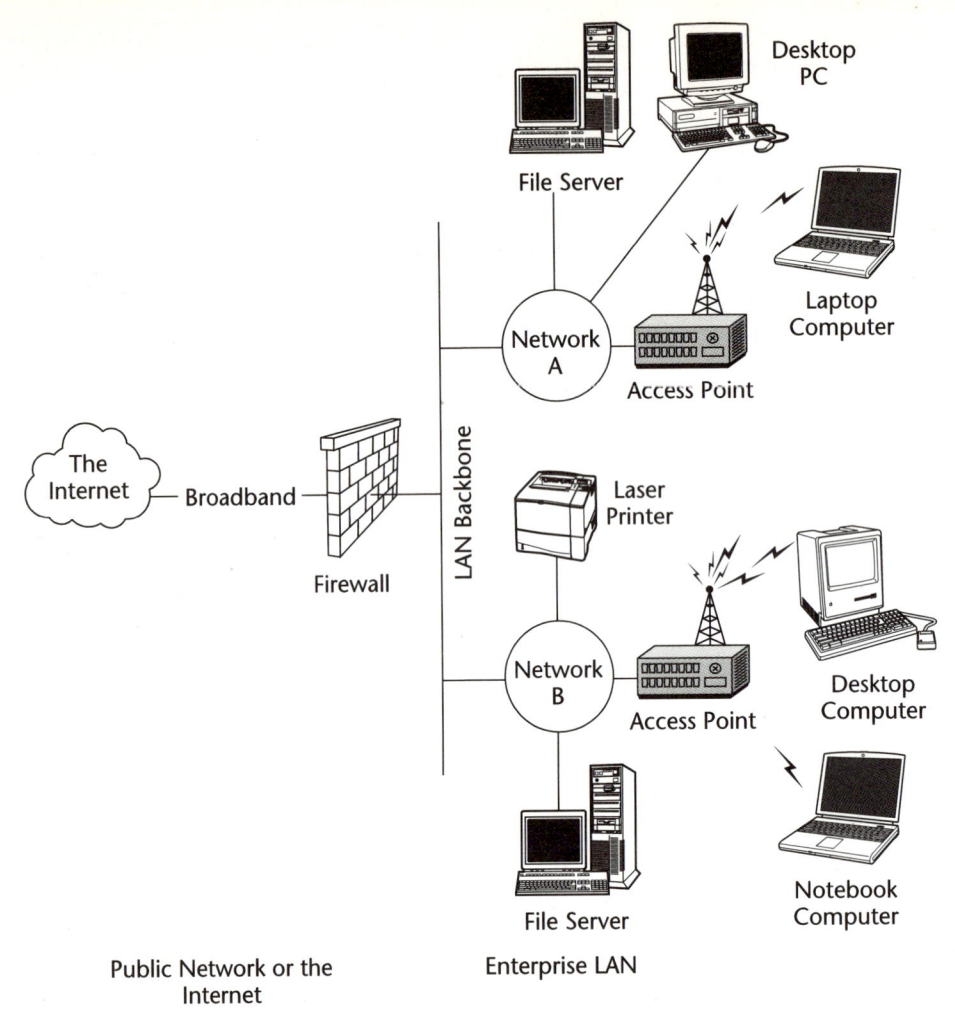

Figure 4.2 Enterprise wireless LAN setup.

The Wireless ISPs come in two flavors:

- Those providing 802.11b-based services at public access points in the urban areas
- Those providing wireless services to customer premises in remote areas

Wireless Access in Urban Public Areas

There are several operators offering high-speed Internet connections at public locations such as coffee shops, airports, hotels, and neighborhoods. These organizations are community-based, providing anonymous and free access, as well as commercial companies that provide such service at cost. One such wireless operator is T-Mobile USA, Inc (for more information, go to their Web site at www.t-mobile.com/hotspot), which has access points at over 1,200 locations across the country including almost all Starbucks in Manhattan and the San Francisco Bay area. Let's look at some of the WISPs that are currently providing service in the different parts of the United States.

Commercial Operators

There are many commercial operators providing Internet services at restaurants, hotels, malls, and other such locations where a large number of people are likely to congregate. Most of the commercial providers are local companies and provide access in the local area with limited coverage areas. One operator, www.hereUare.com is starting a program that allows users to roam between different areas and use services available from a variety of providers through their partnership agreements. Currently there are two main commercial operators providing public 802.11b-based high-speed Internet services. Generally their network access points are connected to T1 or DSL connections to the Internet. They typically have several types of access accounts both that have monthly charges as well as pay-as-you-go plans.

- **T-Mobile USA, Inc. (www.t-mobile.com/hotspot).** T-Mobile by far has the largest network with over 1,200 access points, at the time of this writing, across the country. The service is known as T-Mobile HotSpot Service. T-Mobile HotSpot users need an account to access the T-Mobile HotSpot services, which are available in over 1,200 locations including hotels and airports.
- **Wayport Inc. (www.wayport.com).** Wayport generally caters to business users. They manage access points at many airports and more than 420 hotels, including Four Seasons Hotels & Resorts, Wyndham Hotels & Resorts, Sonesta Hotels & Resorts, Radisson Hotels, and Ramada Inns.
- **hereUare Communications (www.hereuare.com).** According to their Web site, hereUare Communications claims that "unified Wireless Access technology provides the common glue between the myriad of Service Providers, hardware vendors, and wireless Internet access points."

802.11 Public Access Wireless LANs

Most public access wireless LANs are generally managed by community-based independent operators that provide Internet access to the public without any charge. Most broadband providers frown upon such services because they see the public access wireless LANs as a dilution of their market in these service areas. Some of the broadband providers actively monitor the bandwidth usages of its users and at times terminate their service. Nevertheless these public access wireless networks are gaining popularity and are cropping up everywhere. The following is a list of some providers and Internet sites that list wireless LANs:

WLANA (www.wlana.org). Lists various equipment vendors, network software providers, and WISPs.

WiFinder (www.wifinder.com). WiFinder lets users search for a public wireless access point anywhere in the United States.

Wireless Service to Customer Premises

Several ISPs in remote rural areas provide Internet access services via 802.11b to customer premises. These services normally use technologies involving equipment that works in the line of sight. The ISP generally provides a box that can "see" the ISP tower. These boxes communicate with the ISP tower and connect the customer LAN with the Internet through the ISP infrastructure. One such ISP is in Maine. Midcoast Internet Solutions (MIS) started in 1995 in a basement in Owl's Head, Maine. MIS put BreezeNET brand devices on a tower at a high point near Owl's Head, and its new business began. MIS uses a variety of BreezeNET devices:

- Client devices called station adapters (SA) that plug into Ethernet LANs, more or less standard access points (AP)
- Wireless bridges (WB), which connect repeater stations with MISs Internet feed

In a typical end-user installation, MIS brings out a station adapter and an antenna and performs all the wiring necessary to bring an Ethernet connection to the right drop spot. The company sites new locations with either a view to an AP on one of its towers or mountain sites, or by pointing at businesses that host repeating stations.

Costs Associated with Wireless LANs

The popularity of wireless LANs is making it attractive for hardware vendors to manufacture wireless LAN hardware. This popularity is not only bringing thousands of devices to the market, but it is also bringing the prices down. The cost of ownership of a wireless LAN depends on the deployment scenario, the number of users, and the quality of service desired.

SoHo

SoHo deployments are the simplest. These deployments normally require an AP, and a wireless LAN network interface card (NIC) for each computer or device that connects to the wireless LAN. A wireless LAN that uses 802.11b technology with four computers and an AP with broadband connectivity can be built for under $500. An AP without the broadband connectivity might result in even less cost.

Enterprise

Enterprises are the hardest when estimating costs. Enterprise costs depend on the number of users, area of coverage, and the number of APs that might be needed. When calculating the cost of wireless LANs for enterprises, you should be especially careful about the reliability and security of such networks. The best quality components with the highest level of security available should be used to build such LANs. If roaming is desired (which enables the users to roam within a network of two or more APs), when selecting APs, ensure that the APs you purchase support roaming features. If a wireless LAN is to be supported across buildings, then high-power line-of-sight equipment can be used for interconnecting the LANs in two buildings.

Mixing 802.11a with 802.11b devices might also be a solution when interference is a concern. 802.11a can be used to interconnect LANs or to provide a wireless backbone.

WISPs

Most WISP networks today only support 802.11b devices at the point of service. These locations normally include coffee shops, airports, and shopping malls. The cost to build one such site using 802.11b to support up to 10 users would include an AP, a computer to authenticate the users and ensure proper billing, and a broadband Internet connection. A WISP can bring the Internet connection to the point of service using a wireless solution, or using the local communications provider. So, the major cost when setting up a WISP site is the Internet connection.

Deployment Issues

Deployment issues for wireless LANs include the location of the AP, interference with other wireless devices, and network bandwidth.

SoHo

The most common problem in deploying a SoHo wireless LAN is locating the perfect site for the APs. The AP location defines how strong a signal users receive when using the wireless LANs. If an AP is placed at an obstructed location, the network may not perform to its best. APs should be placed where they are least obstructed. A good idea is to perform a site survey and find a location that is central and provides the best signal across the property.

Security of the network is also a concern in SoHo environments. Radio frequency penetrates walls, and if someone on the other side of the wall is aware of a wireless LAN operation, they can easily bring their own wireless LAN cards and connect to the same network. It is, therefore, important that wireless LANs are set up for use with authentication and encryption.

Enterprise

Enterprise wireless LANs have to be very carefully segmented, which means that you must install a good number of APs at a given distance so that there is no interference, and at the same time the APs are not overloaded. A good idea is to first pilot the deployment of a wireless LAN using equipment that seems to fit the need, then experiment using a variety of wireless LAN equipment to measure the throughput the users get, and establish the maximum number of users that should be using a given AP. Minimizing dead zones and high throughput should be the primary concerns when looking at the performance of a network.

The security of the network in an enterprise deployment is perhaps the most important of all. Enterprise networks must be secure, period. To ensure security, make sure that the network always operates in encrypted mode, the shared keys are renewed often, and LAN configuration passwords are kept secret. Most APs and wireless LAN adapters come with WEP-based security. Part two of this book discusses the security requirements and available options in more detail.

Many APs need to provide seamless roaming. The roaming may be required on the same floor, different floors, or among buildings. If a wireless LAN spans across many buildings, rooftop antennas can be used for higher bandwidth. Between floors, if there is existing wiring, it could be used to connect APs.

WISPs

A steady connection to the Internet service, authentication, and correct billing are the three primary concerns for a WISP. WISPs must make sure that the Internet connection always stays up, as it can send customers away if they get a slow Internet connection or if the connection is unreliable. Without authentication, anyone at a WISP location can access the WISP services without the WISP's authorization. This could hurt the WISP's business. Proper software and/or hardware authentication mechanisms must be installed to ensure that only those customers with valid accounts can use the service. Billing must be accurate. If underbilled, it would cost WISP the business; if overbilled, it would cost WISP the customer who was overcharged.

Security

Security is the most debated topic in the wireless LAN community. Wireless LANs can expose secret corporate data and resources to hackers. An unprotected network may also provide outsiders free access to its broadband access. There is a parasitic activity commonly referred to as war driving, which hackers engage in, where the primary purpose is to use the Internet services of other individuals and corporations. War driving is an adaptation of another activity known as war dialing: War dialers use brute force to dial every phone number looking for modems, trying to break into systems and networks. A war driver generally roams neighborhoods, office parks, and industrial areas looking for unprotected networks and sometimes sharing this information on the Internet. To protect a wireless LAN from hackers and other adversaries, it should always be operated in encrypted and authenticated mode.

Health Concerns

All RF devices radiate electromagnetic energy. The health of any living being may be affected by such waves. A good device provides the lowest possible hazard. The 802 standards follow the FCC-mandated radiated power limits.

Most devices sold in the United States come with an FCC identification. If a device comes with an FCC ID, information regarding emission disclosure and frequency usage can be obtained from the FCC Web site at: www.fcc.gov/oet/fccid by providing the FCC ID of the device.

Summary

Wireless LANs are easy-to-deploy networks. The complexity of wireless LANs grows with the number of users that use a wireless LAN and a geographic area that a LAN covers. Wireless LANs can be deployed from homes to large enterprises. Throughput and security are the major inhibitors of a wireless LAN.

PART TWO

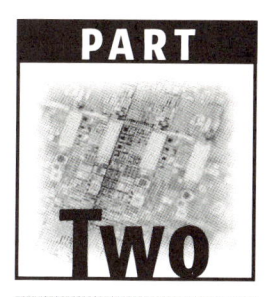

Secure Wireless LANs

LANs are primarily used to share data and exchange information. In business environments, the data often contains critical information on sales, marketing, top-secret deals, and other business-related subjects. At home, it could be important documents, wills, and precious pictures. With increasing dependence on email, electronic documents, and the Internet, it has become a mission-critical task to ensure LAN security. LAN security involves adopting proactive measures to protect and safeguard a LAN from adversaries who may want to gain access to the LAN data, degrade a LAN performance, or make a LAN unusable.

The wireless nature and the use of radio frequency in wireless LANs makes securing wireless LANs far more challenging than securing a wired LAN. Today, wireless LANs have become one of the most interesting targets for hackers. There have been numerous attacks on wireless LANs resulting in widespread skepticism among wireless LAN critics. In order to successfully deploy wireless LANs, you must understand the basic security needs of a wired LAN and that of the wireless LANs. You must carefully choose and deploy appropriate security measures to ensure that the data in the LAN is secured and remains unharmed from attacks that may originate from external and internal network sources. Part 2 of this book talks about all these issues by first walking you through the basics of general wired LAN security, and then talks about the issues surrounding wireless LAN security.

Chapter 5 looks at the basics of network security by discussing the different types of network security, commonly known attacks against computer networks, and the most common practices that are used to ensure security of a LAN.

Chapter 6 examines the special security requirements of a wireless LAN. It provides a brief overview of security primitives in the IEEE 802.11 standard. We explore the weaknesses in the current security model that 802.11 standard-compliant devices use. We also talk about the additional security measures that can be used in 802.11 standard based LANs to provide a higher level of security than defined in the standard.

When you finish reading Part 2, you will understand the basics of securing a LAN that applies to both wired and wireless LANs. You will be able to understand the basic IEEE 802.11 wireless LAN security procedure and the methods that can be used to secure an 802.11 wireless LAN.

# CHAPTER 5

Network Security

Ever since the possibility of remote computer access became available, the temptation for unauthorized access to data and resources has been a painful reality. Computers are continuously being hacked into by malicious or mischievous contenders wishing access to data for any number of reasons ranging from curious exploration to malignant and/or wanton destruction to illegal personal gain. Hackers have used any and all means available to them including trying to connect to computers using dialup connections as well as network connections over the Internet (in this chapter and rest of the book, the term hacker refers to an individual who attempts to gain unauthorized access to a network with malicious intent). In addition to the data, access to the host bandwidth is the prize of the parasite. There is a rich body of hacking information and software codes freely and easily available on the Internet as well as underground hacking networks that detail every known vulnerability in every system. Even a casual and untrained aspirant who is sufficiently foolhardy and resourceful can easily exploit a networked computer using this information without much specialized training. Although there are many laws on the books making unauthorized computer access a serious crime, the seemingly anonymous nature of the Internet combined with the global reach and sheer number of potential targets has made these crimes very prevalent. To further exacerbate the matter, multijurisdictional issues as well as varying treatments of computer laws make prosecution well-nigh impossible except in very-high-profile cases.

Network security has two basic types: network operational security and network data security. Network operational security is concerned with safeguarding, securing, and ensuring a flawless operation of a computer network. Network operational security assumes the roles of information assurance, personnel access control security (controlling who can access the network), defining authorization roles (restricts who can do what on a network), and physical security of the network equipment. Network data security deals with three main areas: confidentiality, integrity, and availability. Confidentiality means that only those who have rightful access should be able to use the information and resources. Integrity implies that only those who are authorized can modify the information. Availability requires that those who need information and resources should be able to access them when they need them.

In this chapter, we explain network operational security, data security, and transmission security. We discuss various aspects of network operational security including access control, physical security, external connection of a network with public networks, and prevalent operational security measures. We talk about the network data security and the basic issues it addresses including confidentiality and integrity and their vulnerabilities. We also discuss the vulnerability of data while in transit (when data is traveling within a network, between two networks, or over the Internet), commonly known attacks, and the measures that can prevent them.

Network Operational Security

Network operational security ensures that a given network is equipped with best-known and appropriate measures to guarantee a reliable, safe (ensures that precious data within the network is never compromised), and intrusion-free (free from the possibility of unauthorized access by intruders or hackers) network. It makes sure that the network is well guarded against malicious attacks and intentions to intrude on the privacy and safety of the network, both from adversaries who are not authorized to access the network and from those who are authorized users of the computer network. To allow a trouble-free operation, operational security includes proactive measures for setting up policies that define how physical access to networking devices will be restricted. It defines and restricts access to the network based on identity (does not allow network access to an individual without proof of their identity) using network access control or authentication, and controls how the network is connected to the Internet or to another network.

The purpose of network security is to prevent and detect unauthorized use of computing and network resources. Prevention measures need to be developed so that unauthorized users can be prevented from accessing part of the computer network they are not allowed to. Detection is necessary in determining attempted and successful network breaches and identifying the systems and the data that have been compromised. Network security is necessary not only to protect the data from unauthorized access but also to protect an unauthorized user from initiating fraudulent transactions under false pretenses such as forged emails or financial transactions.

To adequately secure a network, we need to have a comprehensive plan. In formulating such a plan, we need to consider physical security as well as network authentication and access control; user rights; and user access to workstations, servers, disk space, and printers. In this section we talk about the security issues relating to LAN resources that affect both local and remote LAN users. We talk about physical security, network authentication and access control, common attacks on networks, and ways to ensure operational security in a wired LAN environment.

Physical Security

Physical network security deals with securing physical computing assets and resources from the adversaries. Most common physical security issues include theft and network hacking through penetrating into the physical network cable.

To protect wired networks from theft, in most cases, a well-controlled premises entry system with safeguards against intrusion is necessary. This normally includes a safe environment where computers and networks are located in a hazard-free environment. This hazard-free and safe environment must be premises onto which only authorized personnel are admitted. Network cabling needs to be secured through impenetrable conduits. All connections and network jacks need to be monitored regularly and unused jacks disabled. Servers, routers, and network communication equipment should be located in areas only accessible by authorized personnel. A well-documented chain of custody must be maintained for servers with sensitive data. Central networking resources, such as servers, routers, and network communication resources, should be supplied with conditioned and redundant power systems such as using surge protectors and uninterruptible power supply (UPS) to protect against power-related problems such as surges, blackouts, and brownouts that can cause physical damage and harm electrical components. Data should also be backed up on a regular basis, and offsite data storage must be maintained. Comprehensive disaster recovery plans should be developed, and regular disaster recovery drills must be conducted.

Network Authentication and Access Control

In most cases, the first entry point to a network is through a user workstation. The mechanism of ensuring that a rightful user is accessing the network by validating the authenticity of a user is commonly known as authentication or login. Login is a process that identifies the authenticity of a user based on the credentials he or she provides (for example, username and password). Upon successful login, the user is granted access to the network resources (for example, file servers and printers). Preventing unauthorized access to a network is of primary importance when discussing LAN security. Figure 5.1 shows an example of a network login under Windows 2000.

In most LANs, the user workstations are installed with operating systems (OS) with various levels of built-in authentication features. Most computers allow multiple users to log in and use the system resources. Depending on the OS, the user may log in locally (physically connected to the network), or remotely (for example, connected over the Internet) by authenticating over the network. In either case, the user who wants to access the workstation must be preauthorized to log in. The users are authenticated via a central server called a login server. Each user authorized to access a network must have an account on this login server. The network administrator usually creates these accounts. The privileges and authorization levels are granted to each user when a user account is created. In LAN terms, a given "privilege" normally relates to the type of access a user has over network administration (for example, user account management), whereas authorization refers to a set of permissions that a user is granted to use network services (for example, authorization to access an internal human resources database). Privileged logins, commonly referred to as root or administrator users, should be limited to a small number of authorized users. Access to resources should be mapped through groups of users aggregated in logical collections. For example, in an enterprise setting, users from accounting should belong to a group consisting only of employees working in the accounting department and resources like accounting servers should be restricted to that group. User authentication information is stored in many different ways, which varies in each operating system. However, the standard that is gaining popularity in both the UNIX and Windows 2000 environments is known as lightweight directory access protocol (LDAP). LDAP is a TCP/IP-based protocol used to access user information stored in a specialized database known as an LDAP directory. This directory contains the information necessary to validate the authenticity of a network user. LDAP is supported on Windows 2000, but Windows XP is based on LDAP. In this section, we talk about individual network user authentication, user groups, authentication servers and access control lists (ACLs), and remote user authentication.

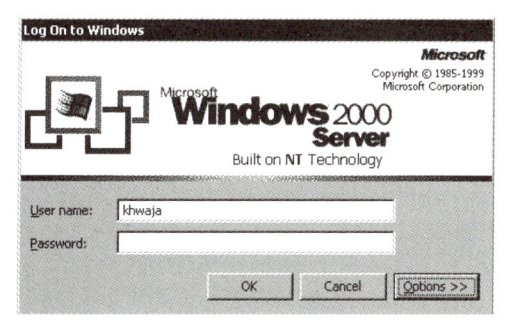

Figure 5.1 Network authentication using user name and password.

Network User Authentication

The most commonly used mechanism for validating the identity of a user from a known authoritative source is called authentication. Network user authentication is used to ensure that only those personnel who are duly authorized can access network resources. Typically, to be authenticated, the user is presented with a screen that collects multiple pieces of information, some of which are well known to all users of the system (for example, a username or login) and some of which are known only to that particular user (for example, a password or a secret word). Generally, a username (login name or screen name) would be known to all participating in a network, and a password that is only known by that user is also required in such a screen. Figure 5.2 shows a network authentication dialog that requires a user to enter username and password. This is known as single factor authentication because it has only one component (password) private to the user.

Normally the authentication information is communicated from the user workstation to the server in a secure manner. For example, Microsoft Windows 2000 uses a challenge-response mechanism in which the server first issues a challenge to the user—for example, asking for information such as username and password—and the user has to provide the correct response to the challenge. In most systems, the passwords are kept on the server in an encrypted format. Figure 5.2 shows a generic network authentication process. The client computers typically collect the password in human readable form known as cleartext and present it to the server in an encrypted form (see *Network Data Security*, later in this chapter, for more information on encryption). Whenever the user requests authentication, the server matches the encrypted password with the one stored in the password database. Depending on the security needs and the operating system, there may be several levels of passwords that are requested by the server before a user is allowed to access a resource.

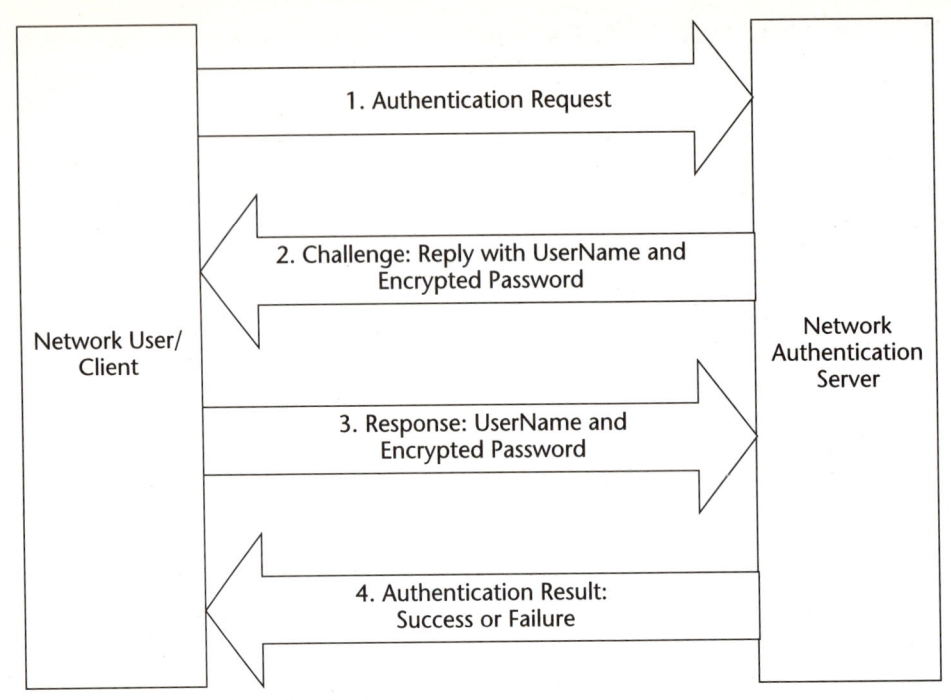

Figure 5.2 Network authentication process.

Although the username and password combination remains the most widely used method of authentication, other means of authentication such as biometric (for example, retina scan or fingerprint) or hardware-based strong cryptographic tokens (for example, smart cards) are being used in scenarios where a higher level of network security is desired. The authentication mechanisms that require more information than just username and password are called n-factor authentication, where n is the number of additional pieces of information that is required to log in. For example, if besides the username and password a retinal scan were also required, it is called a two-factor authentication.

User Groups

In most network deployment scenarios the number of network users directly depends upon the number of personnel in an organization; they do not normally all perform the same job task, nor does everyone manage the network operation. For example, a computer network in an accounting firm with 100 employees may have 60 accountants, 20 administrative support personnel,

10 executives, 5 facility coordinators, and a 5-person information technology (IT) department. Each set of users may need a different set of services—for example, accountants may need access to accounting software and email, executives to confidential data, and IT to the entire network to be able to manage it. To manage and secure access to a given set of services to a set of users is a common construct in security schemes known as user groups. Generally, a user group consists of a collection of one or more users with a unique identifier or name known as a group name. Often users are grouped on the basis of their job function or role within the network environment, and they are assigned appropriate permission to access various network resources. For example, all the users in accounting might belong to a group called accounting, likewise a group to which all users in the facility department belong may be called facilities, and computer systems administrators may belong to a group called sysadmin with permission to access all systems except the servers that contain confidential trade secrets and those containing human resource information. Figure 5.3 shows users and group management under Windows 2000.

In some systems, user groups can contain other groups, resulting in a hierarchy—for example, accountants who deal with clients in Europe may belong to a group known as eu-accountants as a subgroup of accountants. Figure 5.4 shows an example of hierarchical user groups.

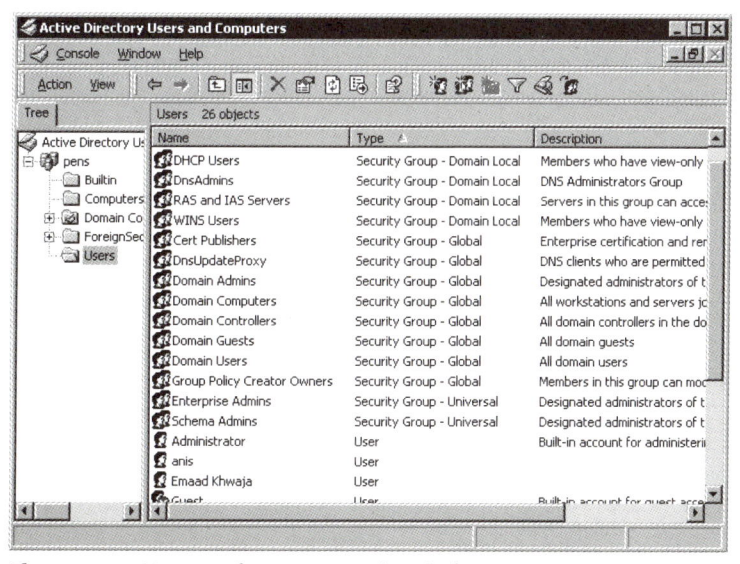

Figure 5.3 Users and user groups in Windows 2000.

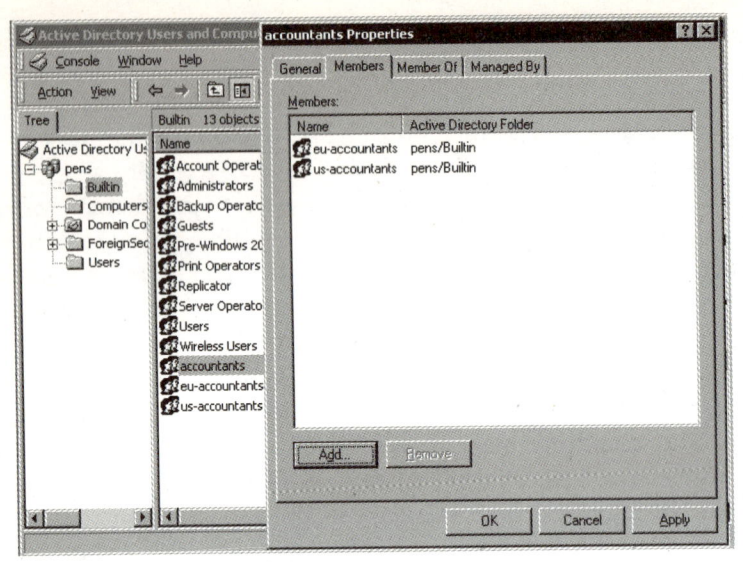

Figure 5.4 Hierarchical user groups.

In essence, user groups provide a higher level of network security and improved network performance by allowing access to the protected network resources only to users in selected groups.

Authentication Servers and Access Control Lists (ACLs)

Authentication servers are the computers that perform the authentication of all network users who wish to access the network. The authentication servers maintain the list of users, groups, and passwords, and the privileges they have. Figure 5.5 shows an authentication server in an authenticated network. This list is known as an access control list (ACL). Access control lists are kept safe and are only managed by a small number of users who are normally the network administrators.

Besides having an authentication server, each computer on a network may have its own authentication mechanism and ACLs if it wishes to allow other network users to access its resource. For example, a networked computer equipped with a high-performance printer may require authentication from those who want to print so as to reduce the cost that the high-performance printer incurs. Likewise, in the Microsoft Windows operating system, file-sharing is controlled using authentication servers and access control lists to restrict access to authorized users only.

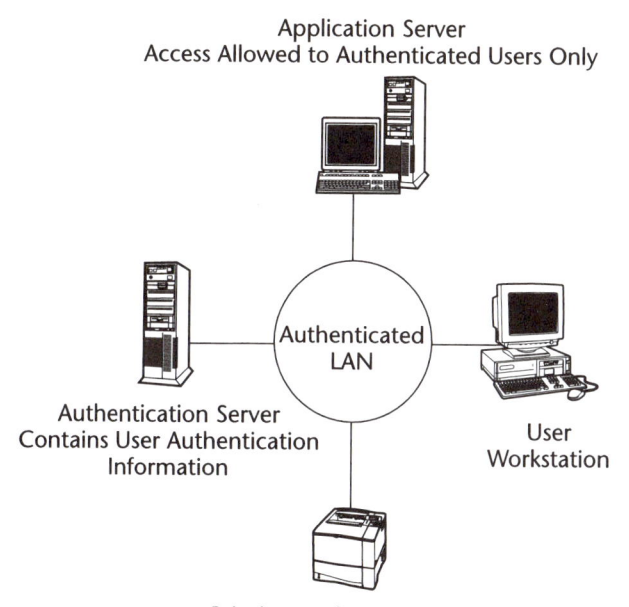

Figure 5.5 Authentication server in a network.

Remote User Authentication

If network users are not present onsite where the physical computer network exists and these users are provided access to the network from remote sites (for example, client site, or from home), then extra security measures are needed to allow users to remotely and securely log on to a network. Onsite users are said to be operating in a trusted environment because they are directly connected to the network. Figure 5.6 shows a remote user connected to a LAN using a dialup connection. Remote users typically access the network through unsecured channels (for example, phone lines or the Internet) and present higher security risks to the overall network.

Typically, remote users are authenticated using an extra level of security in addition to the username- and password-based authentication. Most remote network users are authenticated using standard network protocols; we talk about some of these protocols later in this chapter.

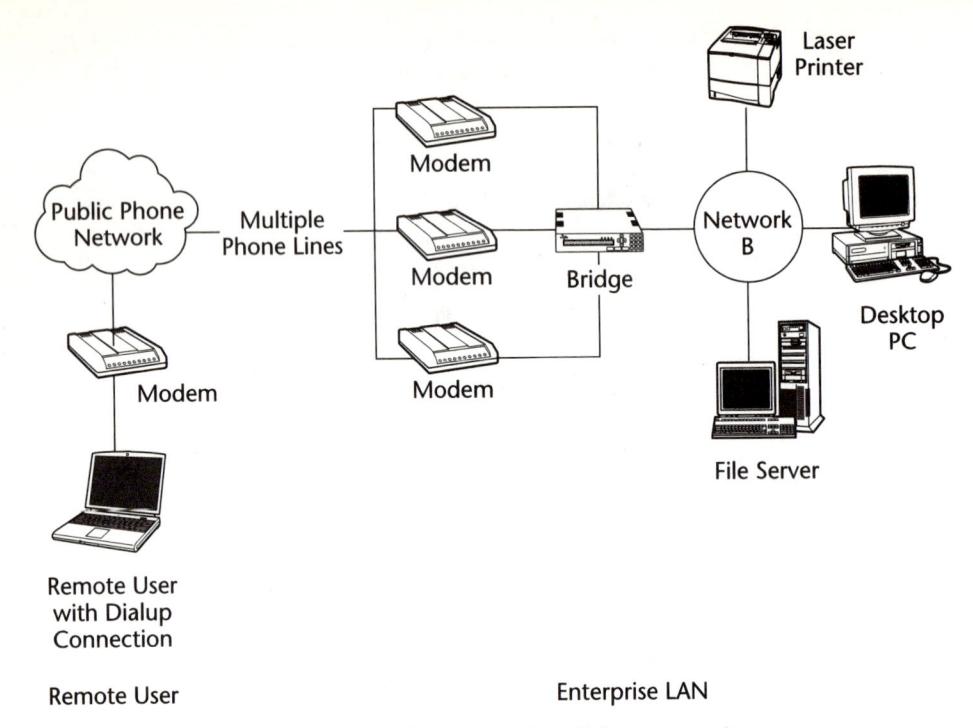

Figure 5.6 Remote user connected to a LAN via a dialup connection.

Common Network Attacks on Operational Security

A network attack on operational security is normally referred to the activities that are aimed to disrupt a network operation, reduce network performance, or completely destroy the network hardware. Though hackers from outside the private LAN perform most network attacks, still attacks from within a LAN are not unheard of either. The attacks that originate from outside the network are called external attacks, whereas those that originate from within a network are called internal attacks.

External Network Attacks

Connecting a network with an external network, especially the Internet, opens up a world of opportunities to internal users, who can benefit from higher connectivity and faster information-sharing, as well as to adversaries who are interested in gaining access to the network for their malicious activities. Just as you are careful about whom you let through the door in your house, a secure network must not allow any unauthorized access to the network. External network attacks are often made possible by insufficient Internet or Extranet security. These attacks are normally conducted by adversaries who cannot gain access to the onsite network hardware and rely on weaknesses in the security that a network uses to protect itself from the outside world. Each type of attack tries to capitalize on a certain weakness that a network suffers. Some of the common external network attacks are password-based attacks, network traffic–based attacks, application- and virus-based attacks, messaging system–based attacks, and operating system–vulnerability attacks.

Password-Based Attacks

As most computer networks use names of persons as usernames for their account identifiers, there is only a limited set of usernames that a hacker has to try when he or she wants to penetrate a network that is protected using the username and password combination. In addition to the username limitations, users choose easy-to-remember passwords that often include names of their significant other, pets, or their social security number; such passwords are easy to guess and add vulnerability to network security. Usernames and passwords usually span a small combination of numbers and letters that can be easily guessed. The vulnerability of username- and password-based authentication systems is further increased by the commonly known conventions for defining the network usernames. Most IT organizations use either the last name of a user or the last name prefixed with the first letter of their first name as their network login name when creating a network account. Password-based attacks capitalize on this limited entropy of usernames and passwords.

Hackers often use a dictionary attack to conduct a password-based attack on a network, where a known set of usernames and passwords are tried against a network login. Another common attack is known as a brute-force attack, in which a hacker attempts all possible combinations of letters and numbers and supplies them to a login screen to log on to a network. For example, in an imaginary network, let's assume that a user Alison Brown is assigned a username abrown and she chooses the word Brooklyn as her network login password, the city she was born in. A hacker finds out that the network on which Alison is a user allows her to log in over the Internet. He or she can try guessing Alison's username by using her first name and the last name. Once a hacker finds out the correct username, he or she can simply use a dictionary attack with the values that might be significant to the geography and language Alison has associations with. He or she then gains unauthorized access to Alison's network.

It is, therefore, important to ensure that users are required to use hard-to-guess passwords. Many organizations require their employees to frequently change their passwords to reduce the risks associated with password-based attacks.

Network Traffic–Based Attacks

Data travels from one computer to another on a network or among networks in small chunks called packets. These packets are normally visible to all computers that have access to the network. Network traffic–based attacks use this vulnerability of networks to intrude privacy and tamper with the information on the network. Common examples of network traffic–based attacks are packet sniffing and denial-of-service (DoS) attacks.

Packet Sniffing

To conduct a packet-sniffing attack, a hacker uses an application program called packet sniffer. A packet sniffer is a program that captures or intercepts data from information packets as they travel over the network. For example, during the authentication phase, a hacker can sniff the data transmitted by a user workstation. The sniffed data in this case may include usernames, passwords, and proprietary information that travel over the network in cleartext. Intruders who gain such information using sniffers can launch widespread attacks on systems by impersonating an authorized user to an authentication server and gaining access to a network that he or she should not have. The packet sniffer problem is further complicated by the fact that installing and using a packet sniffer normally does not necessarily require administrator-level access to a network computer.

Enterprise networks often use advanced authentication mechanisms for remote network authentication and access, which include multifactor authentication and secure authentication servers. Home users, who use digital subscriber line (DSL), cable modems, and dialup connections generally have fewer security primitives available to them than enterprise networks, and are at higher risk. Relative to DSL and traditional dialup users, cable modem users have a higher risk of exposure to packet sniffers as entire neighborhoods of cable modem users are effectively part of the same LAN. A packet sniffer installed on any cable modem user's computer in a neighborhood may be able to capture data transmitted by any other cable modem in the same neighborhood.

Denial of Service (DoS)

Another well-known network traffic–based attack is called a denial-of-service (DoS) attack. This type of attack causes a network computer to crash or to become so busy processing data that you are unable to use it. An example of DoS is an attack by a hacker on a Web site to make it so busy that it cannot handle the Web site lookup by genuine users. In most cases, the latest operating system and computer hardware patches will prevent this attack. The definitive clearinghouse for security-related issues is a federally funded research and development center know as the CERT Coordination Center, or the CERT/CC, operated by the Carnegie Mellon University. CERT/CC was originally called the computer emergency response team. The documents at the CERT/CC site describe denial-of-service attacks in greater detail. For further information, go to their Web site at http://www.cert.org/archive/pdf/DoS_trends.pdf.

Note that in addition to being the target of a DoS attack, it is possible for your computer to be used as a participant in a denial-of-service attack on another system. In such a case a hacker makes a network computer perform an act that causes a DoS attack on a third computer. Attacks of this nature are called application-based attacks.

Application- and Virus-Based Attacks

A hacker normally conducts application- or virus-based attacks by writing computer programs that can affect the performance of a network or an individual computer. These programs are often transported to computers operating in a network—using email, for example—and exploit the weaknesses of a computer operating system to damage data and physical equipment. Examples of such viruses and application programs include Trojan horse viruses and remote network administration programs. Using such applications and viruses, a hacker can also use a naïve computer user's computer to attack other computers or networks, leaving blame on the user.

Trojan Horse Viruses

Trojan horse viruses are a common way for intruders to trick an authorized computer user into installing backdoor programs. These back doors can allow intruders easy access to your computer without your knowledge, change your system configurations, or infect your computer with a computer virus. More information about Trojan horses can be found at: http://www.cert.org/advisories/CA-1999-02.html.

Remote Administration Programs

Many operating systems provide remote management of network resources and identities. Though these are very helpful to computer system administrators, these provide a back door to hackers to gain control over an entire network. For example, on Windows computers, three tools commonly used by intruders to gain remote access to your computer are Back Orifice, Netbus, and SubSeven. These back door or remote administration programs, once installed, allow other people to access and control your computer. Back Orifice is one of the prime examples of such remote administration programs. For more information on Back Orifice, review the following document at CERT Web site: http://www.cert.org/vul_notes/VN-98.07.backorifice.html.

Being an Intermediary for Another Attack

Intruders frequently use compromised computers (those that have been successfully attacked and are under the control of an intruder) as launching pads for attacking other systems. An example of this is how distributed DoS tools are used. The intruders install an agent (frequently through a Trojan horse program) that runs on the compromised computer and awaits further instructions. Then, when a number of agents are running on different computers, a single handler can instruct all of them to launch a DoS attack on another system. Thus, the end target of the attack is not your own computer, but someone else's—your computer is just a convenient tool in a larger attack.

To ensure that a network is secure from such attacks, network users should be discouraged from using programs that are not obtained from a recognized source. Likewise, all users should be requested to report any strange network behavior to the network administrators, and antivirus software should be run on computers participating in a network on a routine basis.

Messaging System–Based Attacks

For a malicious code to be able to execute on a computer in a network, it must first arrive at the computer from the attacker. The easiest mechanism that is available to a hacker is via messaging systems including emails and chat programs.

Email Attachment–Borne Viruses

Viruses and other types of malicious code are often spread as attachments to email messages. Hackers send out emails containing computer viruses to the users on a network that they want to attack. These attachments are normally computer programs that require users to execute them in order to find out the contents of the attachments. It is not enough that the mail originated from an address you recognize. The Melissa virus spread precisely because it originated from a familiar address. Also, malicious code might be distributed in amusing or enticing programs. Many recent viruses use these social engineering techniques to spread.

It is a good idea never to run a program unless you know it to be authored by a person or company that you trust. Also, do not send programs of unknown origin to your friends or coworkers simply because they are amusing—they might contain a Trojan horse program. All inbound and outbound emails should be scanned for viral content, and any email thought to contain a virus should be immediately destroyed.

Email Spoofing or Email Forging

Email spoofing is when an email message appears to have originated from one source when it actually was sent from another source. Email spoofing is often an attempt to trick the user into making a damaging statement or releasing sensitive information (such as passwords). Spoofed email can range from harmless pranks to social engineering ploys. Examples of the latter include email claiming to be from a system administrator requesting users to change their passwords to a specified string and threatening to suspend their account if they do not comply, or email claiming to be from a person in authority requesting users to send them a copy of a password file or other sensitive information.

Note that service providers may occasionally request that you change your password, but they usually will not specify what you should change it to. Also, most legitimate service providers would never ask you to send them any password information via email. If you suspect that you may have received a spoofed email from someone with malicious intent, you should contact your service provider's support personnel immediately.

Internet Chat Programs

Internet chat applications, such as instant messaging applications and Internet Relay Chat (IRC) networks, provide a mechanism for information to be transmitted bidirectionally between computers on the Internet. Chat clients provide groups of individuals with the means to exchange dialog, Web URLs, and in many cases, files of any type. Because many chat clients allow for the exchange of executable code, they present risks similar to those of email clients. As with email clients, care should be taken to limit the chat client's ability to execute downloaded files. As always, you should be wary of exchanging files with unknown parties.

Operating System–Vulnerability Attacks

Besides applications- and network architecture–based attacks, computer operating systems may provide easy point-of-attack to the hackers. These weaknesses are generally features that lack security features.

Unauthenticated File-Sharing

Most networks are equipped with file servers that enable file- and directory-sharing among computer users. File servers are normally equipped with decent security to deter attacks. On the other hand, most individual workstations and computers on a network also provide file-sharing that is normally not secured by network-wide ACLs. These unprotected shared directories are vulnerable to attacks by external users. For example, intruders can exploit unprotected Windows networking shares in an automated way to place tools on large numbers of Windows-based computers attached to the Internet. Because site security on the Internet is interdependent, a compromised computer not only creates problems for the computer's owner, but it is also a threat to other sites on the Internet. The greater immediate risk to the Internet community is the potentially large number of computers attached to the Internet with unprotected Windows networking shares combined with distributed attack tools such as Trojan horse applications.

Web Browser and Mobile Code (Java/JavaScript/ActiveX)

Web browsers have opened up a new arena for hackers and virus developers. A client browsing on the Internet may accidentally execute a program that can have serious negative effects on the computer and the network. There have been reports of problems with mobile code (for example, Java, JavaScript, and ActiveX). These are programming languages that let Web developers write code that is executed by your Web browser. Although the code is generally useful, it can be used by intruders to gather information (such as which Web sites you visit) or to run malicious code on your computer. It is possible to disable Java, JavaScript, and ActiveX in your Web browser. We recommend that you do so if you are browsing Web sites that you are not familiar with or do not trust.

Also be aware of the risks involved in the use of mobile code within email programs. Many email programs use the same code as Web browsers to display HTML. Thus, vulnerabilities that affect Java, JavaScript, and ActiveX are often applicable to email as well as to Web pages.

Hidden File Extensions

Many operating systems use filename extensions to distinguish one type of file from others. Microsoft Windows uses three-letter extensions for identifying a file type. For example, backup.exe could be considered (as filename depicts) an application program that should perform backup operations. Windows operating systems contain an option to "Hide file extensions for known file types." The option is enabled by default, but a user may choose to disable this option in order to have file extensions displayed by Windows. Many email-borne viruses are known to exploit hidden file extensions. The first major attack that took advantage of a hidden file extension was the VBS/LoveLetter worm, which contained an email attachment named "LOVE-LETTER-FOR-YOU.TXT.vbs." When a user first sees this file, he or she thinks that this is a text file and double clicks on the file icon to open the document, but since it is a virus file written in Visual Basic, it starts executing on the user computer and sends emails to all contacts listed in the user's Microsoft Outlook address book.

Securing a Network from External Attacks

Authentication policies must be strongly enforced. Users must be discouraged from sharing passwords with other individuals, and users should be asked to choose passwords that are hard to guess. Antivirus software should be properly installed and run on all computers, and the virus software should be upgraded frequently to prevent attack from new viruses.

When connecting a private LAN to an external network, certain vital computers must be placed in a demilitarized zone (DMZ). A DMZ is that part of the network that is directly connected to an external network or the Internet. Computers in the DMZ are at the highest risk of being hacked into and attacked, so they should be connected to the private LAN through firewalls and routers. Firewalls ensure that only authorized computers in the DMZ or the outer network have access to the private LAN. Firewalls are network devices that do not allow network traffic from outside the network to reach the protected private network. Routers ensure that only traffic addressed to the private network flows from the DMZ to the private LAN. Both firewalls and routers are normally installed such that they monitor both inbound and outbound (from private LAN to the DMZ) network traffic. This ensures that no one from outside can access the computers inside the private LAN and also that no one from inside can engage in activities that are not permitted.

LAN connections to external networks must be provided through a reliable and trusted link. For example, if a LAN is connected to the Internet, the company providing the Internet connection must be trustworthy. The history and security policies of the ISP should be carefully reviewed to ensure that your data would be safe when moving through their infrastructure. The least possible exposure of the private LAN should be allowed. Only those computers that are required to be accessible from the Internet should be exposed.

Internal Network Attacks

Internal network attacks originate from within the network due to malicious intentions or a mistake by a person authorized to access the network. In either case, such attacks should be prevented by properly safeguarding the network resources. Though most of the internal network attacks are authorization-based (improper or unauthorized use of a privilege), most network attacks that can be launched against a network from outside can also be launched from within the network. This means that isolating a network from external networks does not eliminate the possibility of a network attack. File servers and shared disk space, network appliances including printers and external communication systems, network application programs, and databases are often targeted by hackers and adversaries in attacks that originate from within the network.

File Servers and Disk Space Security

The network users normally share files over the network using a central computer called a file server. File servers contain hard-disk drives with capacities to address the needs of the file storage at a given network. The space available to network users on the disk drives is known as disk space. The disk space is secured by dividing the disk into partitions called directories. Access to these directories, where users store their files, is controlled using ACLs to restrict the access to authorized users and groups. Common rights include read, write, execute, modify, and delete. For example, file server "secretfileserver" may contain top secret files that belong to a company—only executives should be allowed access to this server. These servers are normally secured through network security and are only accessible by authorized network users. The most commonly known attacks on the file server are originated either by viruses, which attempt to crash the hard disk by filling it up with garbage information, or by curious internal employees who want to gain information on secret documents that they are not authorized to access.

Network Appliance Security

Network login-based security can be enforced to restrict access to network appliances. Such appliances can include printers, site-entry systems, and network backup devices. For example, only payroll should be able to print to a printer that prints checks. Typically, printers are often shared by attaching them to network servers called print servers. These print servers use network authentication to ensure that a user is authorized to use the printer. Likewise, if a physical entry access system (for example, a building-entry system using key fobs or magnetic swipe cards) is managed using the computer network, it must also be secured.

Application Program Security

Application program security deals with the security that ensures that only designated personnel have access to an application program. For example, only employees dealing with payroll in a given company should have access to an application program that generates or manages the payroll information. An application can work from the OS-supplied security, can implement its own security, or can rely on a database (where it stores its data) to perform security. Application programs that run on a server are specially written to run in authenticated mode because they run on a server on behalf of a remote user.

Users of network application software should be discouraged from sharing passwords with other individuals. In addition, access to network applications should be granted to a minimum number of personnel.

Database Security

Databases provide the data storage for application programs. These databases could contain sensitive information about clients or human resources records that must be kept private. Most databases come with built-in user security with their own username and password authentication schemes. However, since the databases are normally application programs and the data is stored on the disk, the network connection security and the application level security can be applied to databases also.

Network Data Security

One of the basic uses of computer networks is to share data among its users. The information contained in data is often confidential or private. In business environments it could be trade or business secrets, a hidden policy, or classified information. At home, it could be personal emails, pictures, or contracts. All such data must be protected from anyone who should not have access to or knowledge of such information. In a networked environment, such data is vulnerable to be shared or tampered with without your knowledge. For example, let's imagine that Alison, our imaginary naïve computer user, had a very personal file that she did not want anyone to see and she saved it on her computer at work. Since the directory in which she left the file was shared on the Internet, the file was hacked and the very next day she was the talk of the town. Not only are the files residing on a networked computer at risk, but also the data that leaves your computer can be sniffed (seen) by network-monitoring software that has access to the network. One example might be that you sent an email over the Internet to a friend of yours about a multimillion dollar deal that you are engaged in, given that by default all Internet email goes through a number of computers in cleartext (human readable text format). A hacker got hold of the details of the deal, and he or she turned all your dreams into a nightmare by publishing the information on the Internet. This problem is further complicated when remote users connect to a LAN through the Internet. In this scenario, if the data between the remote computer and a user inside the LAN is exchanged in cleartext, all the data transmitted is vulnerable to examination and tampering if it is sniffed by a hacker.

The primary concerns in electronic and network data security are confidentiality and integrity. Where confidentiality means that information can only be accessible by the intended recipients, and integrity means that data cannot be tampered with. Data in a network is vulnerable to both confidentiality and integrity attacks both while it is residing on a computer as well as while in transit between computers on a network or among networks (for example, over the Internet). In this section, we talk about how data is vulnerable to attacks while residing on a computer. We look at the ways the data is secured and briefly discuss the basic cryptographic primitives and how they are generally used to protect and secure network data.

Resident-Data or File Security

Sensitive data residing on a computer's hard disk or on a file server is vulnerable to both confidentiality and data integrity attacks. An adversary can look at the data, gaining information that he or she should not have, or alter the data so that it does not carry the exact meaning it should. An example of such vulnerability would be a file containing secret contract information residing on a network file server that is read by an authorized or unauthorized user. Notice that the file system and network operational security alone cannot meet this vulnerability issue as a user, though authorized to access a particular folder, should not have the ability to read the file.

Protecting Data Using Cryptographic Primitives

The Merriam-Webster Collegiate Dictionary (online version available at http://www.m-w.com) defines the word cryptography as "the enciphering and deciphering of messages in secret code or cipher." Cryptography is the mathematical discipline that is used for keeping information secret and guaranteeing integrity. The most basic cryptographic primitives include encryption (encipherment) and decryption (decipherment). Cryptography has been used for centuries for protecting data confidentiality and integrity. A classic example for the cryptographic procedure is Caesar cipher, known to have been used by the Roman emperor himself for sending messages to his army. In modern days, cryptography is used to protect electronic data from attacks that can damage its confidentiality and integrity. In this section, we look at the fundamentals of encryption and decryption mechanisms and talk about some basic techniques that use cryptographic mechanisms to ensure network data security.

Data Encryption and Decryption

Data encryption is the technique by which known data (that is, plaintext) is transformed into garbled data by using a cryptographic primitive commonly known as a cipher. Substitution ciphers are the simplest ciphers. In substitution ciphers, each letter of the alphabet is substituted by another letter. For example, let's assume that our original message was APPLE; we substitute all occurrences of letter A with letter K, P with Z, L with O, and E with T, then our substitution cipher would work as shown in Figure 5.7.

```
Original message:        APPLE

Using Substitution Table:

    Original Alphabet       Replacement Alphabet
            A                       K
            P                       Z
            L                       O
            E                       T

After substitution we have:

Encrypted message:       KZZOT
```

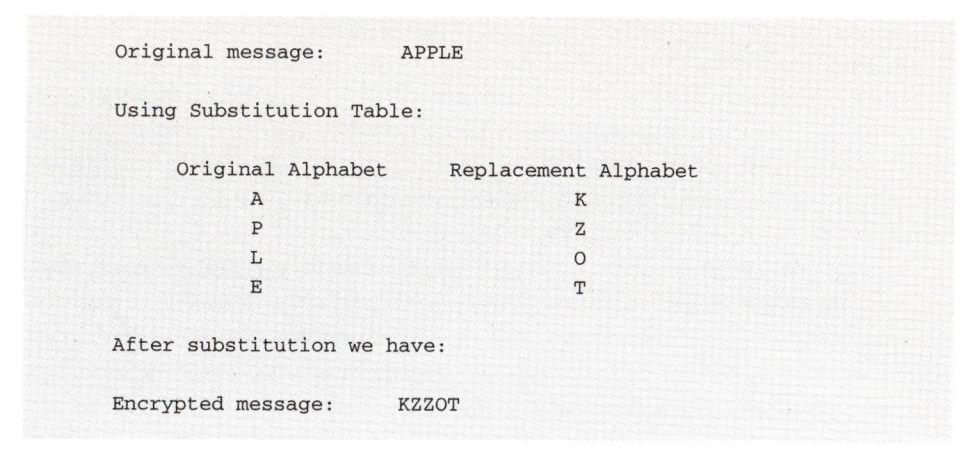

Figure 5.7 Message encryption using a substitution cipher.

Caesar cipher is one of the oldest substitution cipher techniques ever used. In Caesar cipher, the text to be secured is encrypted by replacing each letter of the message with the third letter to its right. For example, A is replaced with D, E replaces B, and Z is replaced with C.

Decryption is the process that enables one to recover the original message from a message that was previously encrypted. To recover the original message APPLE in our example, we need the encrypted message KZZOT and the table that was used to encrypt the message. The decryption in substitution ciphers is the reverse of the encryption process. Let's recover the original message by substituting K with A, the two Zs with two Ps, O with L, and T with E. The recovered message is shown in Figure 5.8.

```
Encrypted message:       KZZOT

Using Substitution Table:

    Original Alphabet       Replacement Alphabet
            A                       K
            P                       Z
            L                       O
            E                       T

Recovered message:       APPLE
```

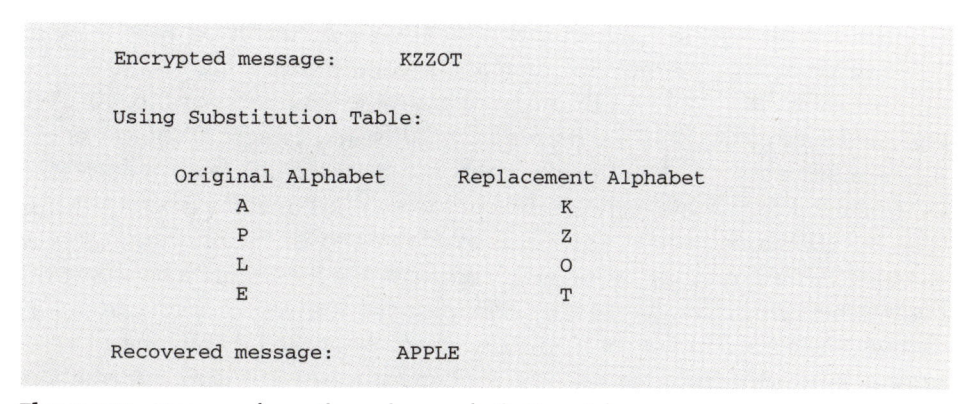

Figure 5.8 Message decryption using a substitution cipher.

The procedure used to perform the cryptographic operation is called an algorithm, the original message is called plaintext, and the encrypted message is called ciphertext. The table or characters used to encrypt a cryptographic message is known as the encryption key; likewise, the key that is used to decrypt a cryptographic message is called a decryption key. In our substitution ciphers, both the decryption and the encryption keys were the table used to substitute the letters in the original message.

Though substitution ciphers are still used for simple message encryption where security is not a concern, most currently used cryptographic algorithms are far more complex than substitution ciphers. There are two types of encryption algorithms: symmetric encryption algorithms and asymmetric algorithms. Symmetric algorithms utilize the same key for both encryption and decryption, whereas asymmetric algorithms employ different keys for encryption and decryption. Examples of symmetric algorithms include Advanced Encryption Standard (AES), Ron's Code 4 (RC4), Data Encrpyption Standard (DES), and Ron's Code 5 (RC5). The most widely used asymmetric algorithms include Rivest, Shamir, Adleman's RSA algorithm, and Whitfield Diffie and Martie Hellman's Diffie-Hellman algorithm.

Network Data Transmission and Link Security

In most network topologies and configurations (for example, Ethernet), when a computer communicates with another over the network, the data travels on the network in cleartext and is available for examination to all computers that share that network path. A computer equipped with sniffing software can be easily used to eavesdrop or alter the transmitted data on a network. This vulnerability has greater impact when a network is connected with another network or to the Internet. In interconnected networks—for example, the Internet—the data might go through several other networks and computers that might attack it, violating its confidentiality and integrity. The confidentiality and integrity of data transferred over a network link is even more crucial when it is transferred between two or more points over the Internet. When data travels through the Internet, it is available to many more computers to examine before it reaches the intended recipient. Purchasing merchandise over the Internet using a credit card is an example of such a case. If credit card information is not transmitted confidentially, it can be hacked by anyone successfully able to sniff the data packets containing the credit card information. Similarly, corporate data transferred or exchanged over the Internet, if not secure, is vulnerable to attack.

To better understand transmission security, let's assume that we have a network A in which Alice is a network user and a network B in which Bob is a user. The two networks are connected with each other using the Internet. We also have Eve, a hacker, who is connected to the Internet using a dialup connection and is able to examine the traffic that flows between networks A and B. This scenario is illustrated in Figure 5.9.

Let's assume that Alice sends Bob a message suggesting that he should send her his credit card info so that she can make reservations for the trip to Hawaii. Bob replies to Alice's message thanking her for her help and includes his credit card information in the message that he sends to Alice. Eve, using her sniffing software, gains access to both messages and therefore is in possession of Bob's credit card information. Eve uses Bob's credit card to arrange the next Hackers Conference on the same day when Alice and Bob plan to be in Hawaii on vacation.

The scenario we just presented shows the vulnerability that messages transmitted as plaintext over the network suffer. Eve was able to understand the messages because they were transmitted without any security, which allowed her to understand the message and successfully exploit the vulnerability of the exchange between Bob and Alice.

There are many different types of attack possible when data is transmitted as plaintext. Our scenario of Bob, Alice, and Eve was the simplest example of transmitted message vulnerability. Just imagine if the entire traffic of two networks was to go through the Internet! A network operating in plaintext over the Internet would be completely insecure no matter how much money was spent to build the security infrastructure.

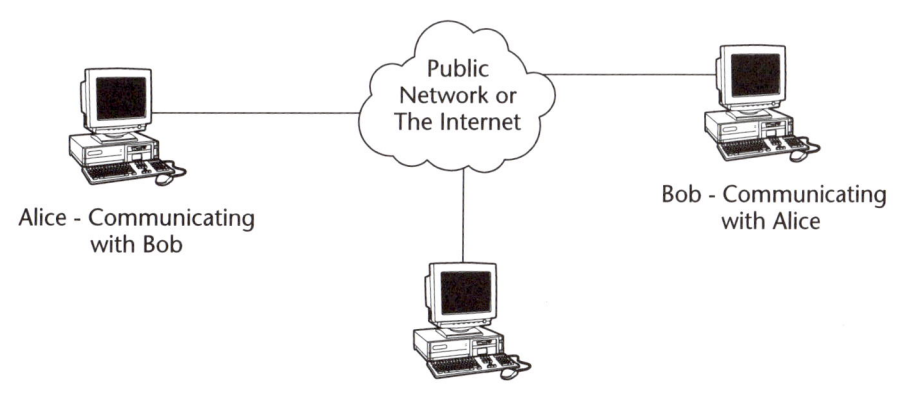

Figure 5.9 Alice, Bob, and Eve in a network attack scenario.

Securing Network Transmission

As shown in the last example, transmitting precious data over the Internet could result in total disaster. In this section, we briefly discuss the common methods used to ensure the confidentiality of data transmitted over the Internet or over a public network of similar nature where data transmitted is subject to the goodwill of other network users. At present, the network transmission is secured by authenticating the users and the devices that communicate with each other to ensure user identity and by encrypting the transmitted data to provide privacy. Whether the communicating entities are two distinct networks or two individual computers, there are two common measures that are used to provide privacy and encryption. Typical methods used to provide network transmission security are hardware link encrypters and virtual private networks (VPNs). In this section, we briefly talk about the authentication protocols that are used to authenticate users and devices, and that link encrypters and VPNs that provide confidentiality over a network.

Authentication over an Insecure Medium

A network communication medium is considered insecure if the users with whom the communication medium is shared cannot be trusted. Primary examples of such media include the Internet and wireless LANs. In this section we briefly talk about significance of authentication in an insecure medium.

Generally speaking, anytime a data packet leaves a workstation, it is assumed to be traveling in an insecure medium, as it is vulnerable to at least eavesdropping. Authentication involves presenting credentials (for example, username and password) to verify one's identity. If these credentials are transmitted over an insecure medium as plaintext, the result is equivalent to presenting your credentials to all the computers in the network (or the Internet, if authentication is performed over the Internet) instead of the entity (for example, the authentication server) you wanted to authenticate yourself to. For example, if user Alice logs on to a corporate network and types her username and password to authenticate herself, her credentials are then transmitted as plaintext over the network. Eve, a hacker working in the same company, can get Alice's username and password by listening to the network traffic from Alice's workstation to the authentication server. Figure 5.10 shows an example of a cleartext authentication attack.

To ensure the security of an authentication mechanism, authentication protocols use cryptographic primitives to add privacy during authentication. Depending on the authentication protocol used, credentials are normally transmitted in encrypted form. Commonly used authentication protocols are: PAP/CHAP, Extensible Authentication Protocol (EAP), and Kerberos.

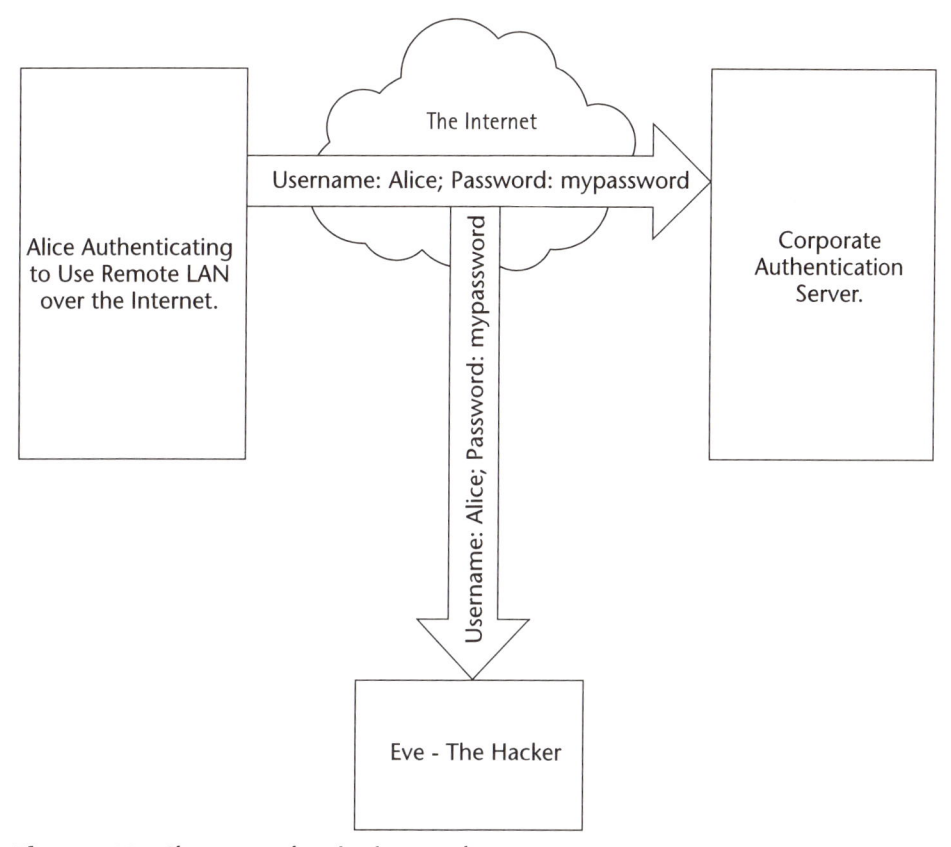

Figure 5.10 Cleartext authentication attack.

Password Authentication Protocol (PAP) and Challenge Handshake Authentication Protocol (CHAP)

Password authentication protocol (PAP) is typically used over a popular link layer protocol known as Point-to-Point Protocol (PPP). PPP is perhaps the most common protocol used to communicate over dialup connect in TCP/IP-based networks such as the Internet or corporate Intranets. PAP is a challenge-response type of authentication mechanism, as described earlier. There are two main weaknesses in PAP: (1) PAP authenticates only the client and not the server; (2) PAP sends passwords in cleartext over the network. Challenge Handshake Authentication Protocol (CHAP) was designed to address these concerns. In CHAP both the client and server authenticate each other using secret words that have been preinstalled in each system. In CHAP all user information including logins and passwords are encrypted. Technically PAP and CHAP authenticate machines talking to each other not the user on the system. Figure 5.11 shows a generic challenge-and-response–based authentication.

Extensible Authentication Protocol (EAP)

Extensible Authentication Protocol (EAP) is a more robust authentication protocol that is used in PPP. EAP supports multiple authentication mechanisms. Unlike CHAP, which does authentication at the onset of the communication at the stage known as Link Control Phase, EAP first sets up the connection using the Link Control Phase but delays the authentication to a later stage known as the Authentication Phase. Doing this allows the authenticator to request additional information, which is used to determine the specific authentication mechanism that should be used in a particular session. This also permits the separation of the communications and the authentication servers. In this scenario the communications server is solely responsible for maintaining the communication link, while all the authentication responsibilities can be delegated to a separate server that performs the actual authentication process. Such a scheme is most commonly used by ISPs, which have thousands of communication servers connected to telephone lines all over the world, but may have only a few central authentication servers.

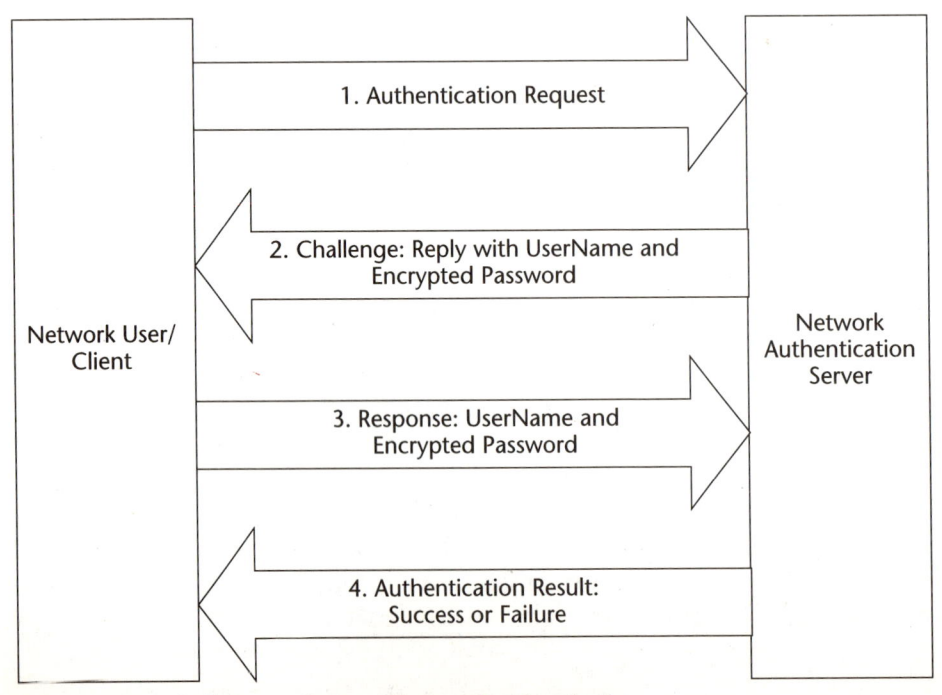

Figure 5.11 Challenge-and-response–based authentication.

Kerberos

Kerberos is a freely available authentication protocol developed and invented by Massachusetts Institute of Technology (MIT) as a solution to network security problems. Strong cryptography is used in Kerberos for both clients and server to prove their identities over insecure network connections. To assure privacy, data integrity, and security, both the client and the server encrypt all transmissions after a client and server have used Kerberos to prove their identity.

Kerberos has been around for a while but has been unsuccessful in gaining a widespread acceptance due to the complexity involved in deploying it in a network environment.

Data Confidentiality over an Insecure Medium

All data that is transmitted over an insecure medium (wired LAN, Internet, or wireless LAN) should be transmitted in encrypted form. Without encryption, any data transmitted over such a medium can be easily compromised, resulting in disasters of varying magnitude. For example, if a macaroni-and-cheese recipe is exchanged over the Internet in plaintext, it might not be too risky, but if two government entities exchange plans about their nuclear warheads over the Internet, the information could be very damaging in the wrong hands.

Insecurity of a medium is, therefore, determined by the type of data that is transmitted over the medium and the parties that might have access to the information while it is being transmitted. Prior to the widespread use of the Internet, most individuals and corporations either used dialup connections or private lines to communicate over long distances. Today, due to the lower cost of using the Internet as a transmission link compared to private lines and dialup connections, most individuals and companies are using the Internet to remotely connect with each other. Wired LANs are considered to be a relatively secure medium, and the use of encryption is reduced to authentication processes only. Sensitive data transmission over the Internet between two entities or two corporate networks interconnected using the Internet is almost always encrypted. The most common way to perform data encryption between two entities includes hardware-level link encrypters and the virtual private networks (VPNs).

Hardware-Level Link Encrypters

Hardware-level link encrypters normally function as ciphers between the transmitting workstations or other network devices. Each end needing to encrypt the data is attached with a link-encrypter hardware device. All data leaving the workstation is encrypted using a cryptographic key that is pre-shared (using a secure medium of choice—for example, over the phone or a personal meeting) between the two entities engaging in encrypted transmission. Figure 5.12 shows the use of link encrypters.

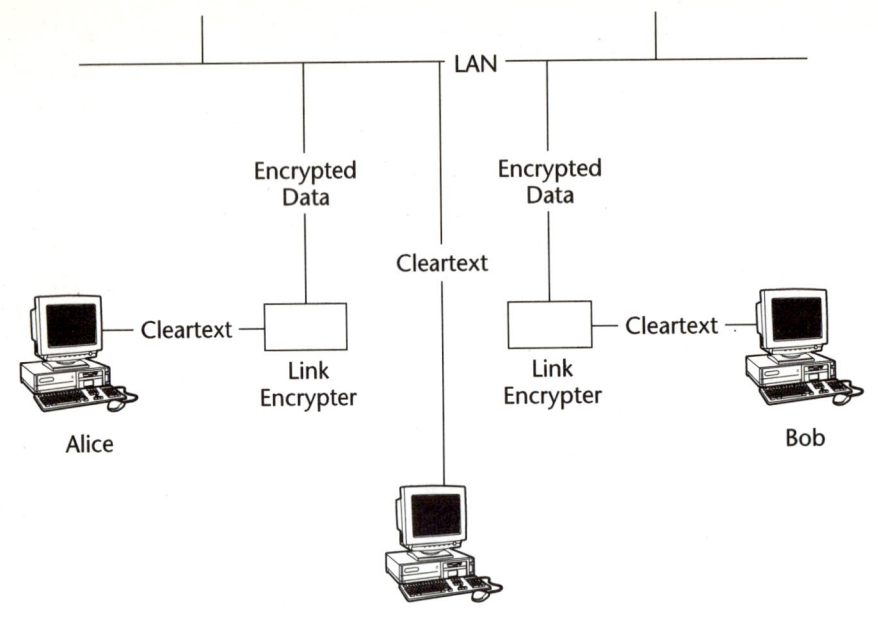

Figure 5.12 Link encrypters securing a communication in a network.

The link encrypters normally use a symmetric algorithm to encrypt the data before transmitting the data over the network. The receiving entity uses the shared key to decrypt the data received and hands it over to the workstation as data received. The wired equivalent privacy (WEP) standard used by the IEEE 802.11 standard uses a flavor of link-level encryption technique to provide confidentiality.

Virtual Private Networks (VPNs)

Today, virtual private networks (VPNs) are the most commonly used means of ensuring confidential data transmission. As the name suggests, VPNs enable an entity (workstation, a computing device, or a remote gateway) to interconnect with a remote network over an insecure medium using a TCP/IP protocol. VPNs provide an encrypted channel to the connected entities, and all the data exchanged between the entities is encrypted. VPNs, therefore, act like a gatekeeper between an insecure network and a LAN where they provide encrypted traffic originating from an authenticated VPN to go through a protected LAN. Most VPNs available in the market today include a robust authentication mechanism that solves the authentication problem as well. VPNs are not only used to enable remote workstations to connect with a LAN over an insecure medium, they can also be used to interconnect two separate networks to form a single virtual private network. In this section, we briefly talk about various components of a VPN, basic operation of VPN involving a VPN gateway and a VPN client over the Internet, and the different types of VPN solutions that are available today.

A Generic VPN Configuration

Components that are required to establish a VPN depend on the deployment and usage scenario. A generic VPN normally consists of a workstation installed with VPN client and TCP/IP network software, a VPN gateway (may consist of software or hardware), and a network link (dialup connection, the Internet, or private line) that connects the VPN client with the VPN gateway (see Figure 5.13).

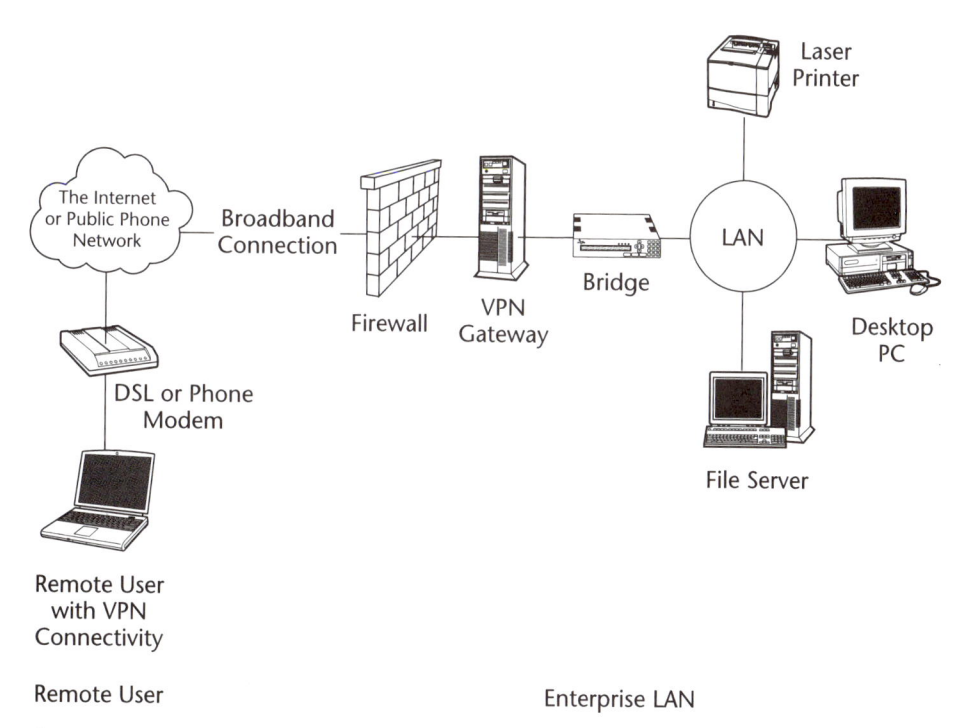

Figure 5.13 VPN connectivity over the Internet.

The VPN client software normally contains cryptographic modules and network software necessary to establish a VPN session with a VPN gateway. A VPN gateway contains cryptographic modules and network software just like the VPN client, but it also contains the VPN authentication database or uses an authentication server for authenticating the VPN clients. VPN gateways are almost always connected with the network that the remote user wants to be connected with upon successful authentication. VPN gateways are often protected with a firewall to avoid DoS attacks.

Basic VPN Operation

Assuming a scenario where a corporate network is equipped with a VPN gateway that is connected with both the Internet and the corporate network, a VPN client installed with VPN client software and connected to the Internet via a dialup connection, a basic VPN operation can be summarized as follows:

- VPN client uses the credentials provided by the VPN user to authenticate him or her to the VPN gateway.
- Upon successful authentication, VPN gateway negotiates a cipher-suite (set of cryptographic algorithms and encryption keys) with the client VPN software in a secure manner.
- VPN gateway assigns an unused IP address to the VPN client. VPN client uses this IP address to identify itself when communicating with the remote network.
- Encrypted session between the client and server begins. Client can now access the network resource behind the VPN gateway as VPN gateway forwards any transmission that it receives from the VPN client to the corporate network it is securing.

Commonly Used VPN Implementations and Protocols

There are many different VPN protocols and implementations available today. The most commonly used VPN implementations are based on two major protocols: Point-to-Point Tunneling Protocol (PPTP) and the Internet Protocol Security (IPSec).

Point-to-Point Tunneling Protocol (PPTP)

The Point-to-Point Tunneling Protocol (PPTP) was sponsored by Microsoft Corporation, and it is implemented in Microsoft Windows 2000, Windows 98, and the newer version of Windows as well as Linux and other operating systems and popular network security equipment such as Cisco and Watchguard. PPTP is basically an extension to PPP and allows one network to be routed over another network to connect to a private IP address space in a VPN environment. For example, a particular organization may use one set of unrouted IP address range 10.168.0.1 to 10.168.0.254 at its main offices but may want its remote employees to connect to it through the Internet using 192.168.0.1 to 192.168.0.254. In such a case the employee would first connect to a local ISP and would be allocated an IP address for that session. This IP address would of course be in the range of addresses serviced by that ISP, and to successfully communicate with the rest of the computers in his or her office, this user needs an IP address in the range of addresses serviced by the head office. Using a PPTP connection, this user will establish a link to the computers in his or her office. PPTP does not use encryption, but a separate encryption protocol may be used with PPTP.

Internet Protocol Security (IPSec)

Internet Protocol Security (IPSec) is an Internet Engineering Task Force (IETF) standard, and it is documented in Request for Comments (RFC) rfc2401 [4]. This protocol is used in a manner similar to PPTP. IPSec is a much more robust protocol and has greater flexibility and features than PPTP. The security levels in IPSec are also very high, and it allows for a variety of cryptographic mechanisms and varying key sizes. Unlike PPTP, the security keys and secret words between the client and servers must be exchanged in advance of making the connection. IPSec is the most widely deployed VPN protocol.

Securing Network Data

To add data security features to a network, all important documents and communication must be encrypted. To ensure network data security, besides practicing a strict operational network security policy, at least the following practices must also be adopted:

- The files placed on a server must always be encrypted. The directories containing the file must require authentication with appropriate permissions. For example, all patent pending documents should be kept on a secure file server in encrypted form with read permissions to few and modify permissions to only the inventor and the legal team.
- All important email messages must be sent in encrypted form. When sent in encrypted form, email messages are safe from eavesdropping and integrity attacks.
- All external network connections must be secured through the use of VPN. This allows only authorized personnel to access the network remotely. In addition to authentication-based security, the encryption feature of VPNs must be used to ensure that transmitted data is not eavesdropped upon or tampered with.

Summary

In wired networks, physical security is much easier to manage. To protect wired networks, in most cases, a well-controlled premises entry system with safeguards against intrusion and cabling is enough. The operational security of a network is maintained by allowing only authorized personnel access to the network. The most common attacks on networks include password-based attack, computer viruses, and messaging system–based attacks. Using cryptographic primitives ensures data confidentiality and network data security. These cryptographic mechanisms use encryption technology and encryption keys to provide privacy and integrity of the data when data travels between computers on a network. To protect a network from hackers, it must contain primitives to ensure operational security as well as data security.

In Chapter 6, we discuss various mechanisms currently being used to secure wireless LANs. We talk about security requirements of wireless LANs, the IEEE 802.11 security architecture, the shortcomings of 802.11 security protocols, the future of 802.11 security, and the basic extensions to 802.11 security that can help overcome the known security weaknesses of the 802.11 security architecture.

CHAPTER 6

Securing the IEEE 802.11 Wireless LANs

Due to the popularity and the ease of use that wireless LANs provide, organizations today are rapidly deploying wireless LANs to provide mobility to their users. Individuals and home users are enjoying the ease of setting up networks, and wireless ISPs are providing services at public places like coffee shops and shopping malls. Unfortunately, most such deployments ignore the basic security issues that are related to the currently available wireless LAN technologies. The main security issue with wireless networks, especially radio-frequency–based networks (for example, 802.11-based networks), is that the wireless networks intentionally radiate data over an area that may exceed the limits of the area the organization physically controls. For instance, 802.11b radio waves at 2.4 GHz easily penetrate building walls and are receivable from the facility's parking lot and possibly a few blocks away. Someone can passively retrieve all of a company's sensitive information by using the same wireless LAN adapter from a distance without being noticed by network security personnel. These vulnerabilities of wireless LANs have made them one of the prime targets of the hacker community today. Security issues surrounding wireless LANs become even more critical when a wireless LAN is connected to the Internet. In this situation, hackers are not only interested in gaining access to a wireless LAN to tamper with it, they are also interested in gaining unauthorized access to the Internet for free high-bandwidth connection and impersonating network users. It is, therefore, extremely important that wireless LAN security and the risks and vulnerabilities are well understood before they are deployed or used.

In this chapter, we examine the security requirements of wireless LANs to ensure secure operation and data transmission, the IEEE 802.11 standard security wired equivalent privacy (WEP) standard, the weaknesses in the 802.11 standard security model, and the measures currently available to improve and build secure wireless LANs using the IEEE 802.11 standard–based technologies.

Wireless LAN Security Requirements

Security of a LAN is often dictated by the physical properties of the medium it uses for communication, the methods used to transmit the data, the protocols that are used to control the security of the data transmitted, and the policies that a LAN enforces to ensure authorized use. For example, private wired LANs are considered secure networks as long as they are not connected to an outside network (for example, the Internet), the LAN equipment and the wiring are physically secured, only authorized personnel are allowed access to the network, and the network security policies are strongly enforced. Wireless LANs use airwaves to transmit the data and are considered inherently insecure because their data transmission medium is not physically bound like their counterpart, the wired LANs. Transmitted over the airwaves, the data in a wireless LAN, which spreads in all directions, allows its users the freedom to move about. However, this also means that adversaries do not require a physical connection to hack into the wireless LAN. Instead, he or she needs to be present in the physical range where radio signals can be intercepted. For example, if a wireless LAN emits a radio signal that reaches up to a radius of one mile, all hackers within the one-mile radius can easily intercept the signal and possibly conduct an attack on the network. A standalone wired LAN (one that is not connected to an outside network) is far more secure when compared with a standalone wireless LAN. Wireless LAN security can be compared to wired LAN security by using the example of old cordless phones that did not securely communicate with their base stations. For example, assume that your neighbor and you both have one of the old cordless phones that did not encrypt the signals between the handset and the base station. Every time you pick up the phone to make a phone call, provided that your and your neighbor's phone were using the same frequency channel, you will be able to eavesdrop on your neighbor's conversation. Wireless LANs are, therefore, inherently insecure and appropriate measures must be taken to ensure a high-performance and secure wireless LAN.

To secure a wireless LAN, both operational security (see Chapter 5, "Network Security") and data security must be enforced. The security issues of wireless LANs are similar to those of the wired LANs, and in this chapter, we discuss only the issues that relate to operational security and the data security issues of the wireless LANs. For more information on wired LAN security, see Chapter 5.

Wireless LAN Operational Security Requirements

Operational security of the wireless LANs deals with the security primitives that provide a flawless operation of a wireless LAN. Operational security must be implemented to avoid any threats that can affect the day-to-day operation of a wireless LAN. Most such threats are possible due to poorly configured wireless LAN setup, the inherent radio frequency–based transmission medium, the technologies and the protocols used to transmit the data, or insufficient user authentication. In this section, we look at the general security requirements that are necessary to ensure the operational security of a wireless LAN. We also examine the need for securing wireless access points (APs), the radio frequency (RF) methods that are used to transmit data over the airwaves, link-level security that allows wireless equipment to operate in a wireless LAN, and wireless LAN authentication. We also talk about the most common known attacks on wireless LANs.

Wireless Access Point (AP) Security

Most wireless LANs operate in infrastructure mode (see Chapter 2, "Wireless LANs") where a wireless access point (AP) coordinates communication among its users by acting as a hub and transmitting data received from one user to another. For example, let's assume a wireless LAN that consists of two users (Alice and Bob) with computers equipped with wireless LAN adapters (along with necessary software and drivers) and an access point. In this example, when user Alice sends a message to user Bob, Alice's wireless LAN adapter transmits the data to the AP, which in turn looks at the data packet that is intended for Bob, and transmits the data to Bob. The use of APs to route all the traffic among its users makes a wireless LAN less reliable, as all the users on a given wireless LAN share the same AP. This may result in a single point of failure, where anything happens to the AP. For example, if an AP gets too busy or it is hacked, it affects the performance of the entire network. In addition to the single-point-of-failure APs, most APs that are available today can be managed using a wireless connection. This management feature, though extremely useful, allows an adversary to attempt to break into the security of an AP and possibly take over its operation.

The number and types of attacks on wireless APs has been growing steadily, and will continue to do so as they become more popular and widespread in deployment. These attacks are easy to launch and some can be difficult to detect on your network via traditional means. The most commonly known attack on an AP is conducted by a wireless LAN adapter that constantly sends messages to an AP, making it so busy that it cannot reply to the messages sent by real users of a network. This attack is known as a denial-of-service (DoS) or flood attack, as the AP is flooded with bad requests from the rogue wireless LAN adapter making the AP too busy to service genuine requests from authorized users. Besides flooding attacks, there are other attacks—for example, AP administration attacks, in which an AP is highjacked by an adversary who then controls all traffic through the AP. In scenarios where an AP connects a wireless LAN to a wired LAN, more advanced attacks can be launched that target the wireless LAN as well as the wired LAN to which the wireless LAN is connected.

Therefore, it is important to use APs that include measures to defeat the known attacks. For example, a secured wireless LAN must contain APs that have built-in authentication mechanisms for authenticating both the network users and the users who are allowed to manage the AP features. Carefully designed APs also contain primitives for securing against DoS. More advanced APs come with a built-in router and a firewall to prevent unauthorized traffic to enter the wireless LAN.

Radio Frequency (RF) Method

The data in a wireless LAN travels over the airwaves by using radio frequency as the carrier. Using radio frequency as the carrier means the transmitting LAN device—for example, a wireless LAN adapter—superimposes the data on a predefined radio frequency and then transmits it over the air. The receiving LAN device separates the data from the carrier wave, converts it into digital signal, and interprets accordingly. The security of the data transmitted over the air can be affected in many ways, some of which include: jamming the radio frequency, which makes a wireless LAN inoperable, and eavesdropping on the authentication of the data, which reveals the user information (the data security in a wireless LAN is discussed later in this chapter). A typical wireless LAN has a range of up to 300 meters per AP. Under most circumstances and depending on the placement of the AP, just like cordless phones, the waves carrying the signals can easily penetrate through the walls. It is, therefore, important that the APs be placed at or near the center of a wireless LAN site to reduce the distance that the airwaves can travel.

The method used to transmit the data over the airwaves is also of prime importance when considering the security of a wireless LAN. There are many different methods used today to transmit the data in a wireless LAN. The most common are direct-sequence spread spectrum (DSSS) and frequency-hopping spread spectrum (FHSS). FHSS is considered more secure and resilient to attacks compared to DSSS. In FHSS, the channel at which data is transmitted keeps switching, whereas in DSSS the data is transmitted at a fixed channel. (For more information on radio frequency methods, see Chapter 2.)

When choosing a wireless technology, it is important to choose a technology that provides the best RF security primitives. The most current available wireless LAN equipment—for example, 802.11-standard devices—utilizes the DSSS method.

Link-Level or Network Adapter Authentication

Many wireless LANs authenticate users based on link-level authentication, in which a network adapter in a wireless LAN communicates with an AP or with another adapter that identifies itself using its media access control (MAC) address. MAC addresses are 48 bits long, expressed as 12-hexadecimal digits (0 to 9, plus A to F, capitalized). These 12-hex digits consist of the first 6 digits (which should match the vendor of the Ethernet interface within the station) and the last 6 digits, which specify the interface serial number for that interface vendor. These addresses are usually written hyphenated by octets (for example, 12-34-56-78-9A-BC). By industry standards, MAC addresses are burnt into and printed on the network adapters used to communicate in a wireless. If configured properly, most wireless LAN APs are designed so that they can authenticate a user based on the MAC identifiers that are preprogrammed in the AP by the administrator. That means that APs let in only those network adapters, and hence users, that identify themselves with known MAC addresses. The MAC-based authentication is considered complex and cumbersome because it requires every AP in a network to have the MAC address of every adapter that might use the AP services. MAC-based authentication is also considered weak because of the availability of LAN adapters that can be reprogrammed to use a different MAC address. In such a case, a hacker acquires a wireless LAN adapter that is programmable and reprograms the adapter to use a MAC address that is known by a network he or she wants to attack. The hacker then conducts an attack by bringing his or her computer equipped with a rouge LAN adapter within the radio range of the AP. The LAN adapter with the forged MAC address leads the AP into believing that it is a previously authorized network adapter and successfully gains access to the LAN.

MAC-based authentication should be used only as a supplementary authentication method. If MAC-based authentication is used, the network becomes vulnerable to such rogue wireless LAN adapters, which may impersonate an authorized wireless LAN adapter to gain access to the network.

Network Authentication

If a communication link is successfully established between two wireless LAN devices (for example, an AP and an adapter), the next step by a user is to establish a network session by authenticating himself or herself to the network (AP or an authentication server that an AP uses). Unfortunately, most currently available wireless LAN technologies do not include a robust mechanism for network authentication. Most network technologies—for example, 802.11-standard devices—only allow a service set identifier (SSID)-based authentication, in which each AP is assigned a unique identifier consisting of letters and numbers and broadcasts this identifier to show its presence. All wireless LAN devices use this identifier to communicate with the AP.

The SSID-based authentication is extremely weak and only provides AP identification. The SSIDs are easily programmable on most APs. An attack on APs, known as rogue AP attack, is the most popular attack that involves an adversary planting an AP in a wireless LAN with the SSID set to the one that is used by the network users. If the network relies only on the SSID of an AP for its authentication, the rogue AP successfully gains access to all the incoming traffic from wireless LAN adapters that is addressed to the intended AP. More information on authentication mechanisms used in 802.11 is provided in *802.11 WEP Authentication,* later in this chapter.

Wireless LAN Data Security

As mentioned in Chapter 5, "Network Security," data in transit in an insecure medium must always be protected using encryption primitives. Encryption-based data security is even more important in wireless LAN, because without encryption the data is available for examination to all authorized users and anyone who can receive the RF signals.

Most attacks on data security in a wireless LAN are conducted by analyzing the LAN traffic. If the data is not transmitted in encrypted form, anyone can easily eavesdrop upon, alter, or damage it. The data security of wireless LANs is further degraded by the fact that most wireless LAN equipment today does not have security features enabled by default. A user has to manually configure the security parameters, which also inhibits the use of encryption in wireless LANs for data security.

The encryption parameters that are important to consider when choosing a wireless technology include the security strength of the encryption technology used to encrypt the transmitted data and the key size that the encryption algorithm uses. It is also important to keep up with the wireless LAN community to learn the new data security threats and the solutions to defeat them.

The Institute of Electrical and Electronics Engineers (IEEE) 802.11 Standard Security

The Institute of Electrical and Electronics Engineers 802.11 standard is the most widely accepted and deployed wireless LAN technology today. (See Chapter 3 for more information on the IEEE 802.11 standard.) The 802.11 standard defines two mechanisms for providing security to the wireless LANs that comply with this standard: service set identifiers (SSIDs), which are used for access control to an AP, and Wired Equivalent Privacy (WEP) protocol intended to provide data security and for over-the-air transmission.

Service Set Identifiers (SSID)

Each AP on a wireless LAN based on the IEEE 802.11 standard is identified with an identifier or name called an SSID. An SSID is a unique identifier of up to 32 characters that is attached to the data sent over a wireless LAN and acts as a password when a wireless LAN device tries to connect to an AP. The implementation of SSID varies among the manufacturers of the 802.11 devices. Some devices allow only one identifier, while others may allow up to four or more SSIDs. The SSID can be used to differentiate one wireless LAN from another, so all access points and all devices attempting to connect to a specific wireless LAN must use the same SSID. A device will not be permitted to join the wireless LAN unless it can provide the unique SSID that is used by the AP. SSID is contained in the radio beacon messages (periodic radio signals) that all APs send out at regular intervals over the air to announce their presence. These beacons are sent in cleartext; since an SSID can be sniffed in plain text from beacon data, it does not supply any security to the network.

Wired Equivalent Privacy (WEP) Protocol

Because wireless is a shared medium, everything that is transmitted or received over a wireless network can be intercepted. To protect the integrity of the data, ensure the privacy and authentication of over-the-air transmission between wireless LAN APs and the wireless LAN adapters, the IEEE 802.11 standard stipulates an optional encryption protocol called Wired Equivalent Privacy (WEP). The goal of adding these security features is to make wireless traffic as secure as wired traffic. The IEEE 802.11 standard provides a mechanism to provide security by encrypting the traffic and authenticating wireless LAN adapters. WEP is the most criticized topic among the wireless LAN critics. Although WEP is optional, support for WEP with 40-bit encryption keys is a requirement for wireless fidelity (Wi-Fi) certification by the Wireless Ethernet

Compatibility Alliance (WECA), an organization set up by wireless LAN equipment manufacturers to ensure interoperability of their products and issue Wi-Fi certificates to all interoperable devices, so WECA members invariably support WEP. It is important to understand the features and vulnerabilities of WEP to decide whether 802.11-standard security is enough or additional security might be desired for a given deployment.

WEP Implementation Details

WEP is implemented at the data-link layer (see Chapter 1, "Networking Basics") of all Wi-Fi-compliant devices to provide an equivalent level of privacy as is ordinarily present with a wired LAN. The WEP protocol provides both privacy and authentication services and consists of an encryption algorithm, a shared-secret key, and an initialization vector. The protocol components and services are described next.

Ron's Code 4 (RC4): The WEP Encryption Algorithm

WEP uses the Ron's Code 4 (RC4) stream cipher as its encryption algorithm that was invented by Ron Rivest of RSA Security, Inc. The RC4 encryption algorithm is a symmetric cipher (an encryption algorithm that uses the same key for both encryption and decryption) that supports a variable-length key. Research has shown that the strength of an encryption technology often depends on its key length. The performance of encryption technology in the WEP protocol was compromised to a lower size key due to the United States export control regulations that did not allow any encryption technology over 40 bits (5 characters long) to be exported outside the United States. To avoid conflicting with United States export controls that were in effect at the time the standard was developed, 40-bit encryption keys were required by IEEE 802.11, though many vendors now support the optional 128-bit (64 characters long) standard.

WEP Shared Key: The WEP Encryption and Authentication Key

RC4 requires the use of a shared symmetric key. The IEEE 802.11 standard provides two schemes for defining the WEP keys to be used on a wireless LAN. With the first scheme, a set of as many as four default keys are shared by all wireless LAN adapters and APs in a wireless subsystem. When a client obtains the default keys, that client can communicate securely with all other stations in the subsystem. The problem with default keys is that when they become widely distributed, they are more likely to be compromised. In the second scheme, each client establishes a "key mapping" relationship with another station. This is a more secure form of operation because fewer stations have the keys, but distributing such unicast keys (keys that are used by only two systems) becomes more difficult as the number of stations increases.

Initialization Vector (IV)

An initialization vector (IV) refers to a set of characters that are randomly generated and are used with shared keys to create the true encryption keys. The shared key remains constant while the IV changes periodically. The IV extends the useful lifetime of the secret key and provides the self-synchronous property of the algorithm. Each new IV results in a new key sequence, thus there is a one-to-one correspondence between the IV and the output. The IV may change as frequently as every message, and since it travels with the message, the receiver will always be able to decrypt any message. Therefore the data of higher layer protocols (for example, IP) are usually highly predictable. An eavesdropper can readily determine portions of the key sequence generated by the (Key, IV) pair. If the same pair is used for successive messages, this effect may reduce the degree of privacy. Changing the IV after each message is a simple method of preserving the effectiveness of WEP.

The WEP Protocol Operation

The WEP algorithm provides both authentication and encryption to 802.11 LAN devices. WEP uses a shared key, and the same key is used to encrypt and decrypt the data. In other words, WEP uses a string of up to eight characters and the same shared key is used by the AP and the wireless LAN adapters. The WEP encryption algorithm works as shown in the steps that follow.

WEP Encryption Procedure for Data Security and Privacy

1. Generate the encryption key generation from the shared key.

    The 40-bit shared key is concatenated with a 24-bit long initialization vector (IV), which is a randomly generated data, resulting in a 64-bit total key size. The resulting key is fed into the RC4 algorithm to create the actual encryption key. Figure 6.1 shows a theoretical key generation using the WEP algorithm.

```
Concatenated-Key = Shared-Key + IV

Encryption-Key = RC4(Concatenated-Key)
```

Figure 6.1 WEP key generation.

```
CRC-Value = CRC32(Original-Message)

Message-with-CRCCheck = Original-Message + CRC-Value

Encrypted-Message = (Message-with-CRCCheck) XOR Encryption-Key

MessageSentToPeer = IV + Encrypted-Message
```

Figure 6.2. Data encryption using WEP.

2. Encrypt the data using encryption key.

   A 32-bit cyclic-redundancy-check (CRC32) operation, an integrity algorithm used to protect against unauthorized data modification (a method used for the detection of errors when data is being transmitted. A CRC is a numeric value computed from the bits in the message to be transmitted. The computed value is appended to the tail of the message prior to transmission, and the receiver then detects the presence of errors in the received message by recomputing a new CRC and compares it with the CRC that is sent with the data), is performed on the data by feeding the data to be encrypted into the CRC algorithm, which results in 4 bytes. The resulting 4 CRC bytes are concatenated to the original message. The resulting sequence is then encrypted using the encryption key generated in Step 1 by performing a mathematical operation called the bit-wise exclusive-or (XOR). Exclusive-or is a mathematical operation that compares the 2 bits at each bit position in two given values, for example value A and value B. If the bit at the specified position is 1 in either value A or value B, but not in both, then that bit will be set to 1 in the result. XOR is often used in symmetric cryptographic algorithms, where data to be encrypted is XORed with an encryption key for encryption; and to recover the original data, encrypted data is XORed with the encryption key. The result is an encrypted message equal in length to the number of data bytes (original data) plus 4 bytes. The final message, the encrypted message, is sent to the peer (that is, from AP to adapter or adapter to AP) with the IV pre-pended to the encrypted message. The encryption step works as shown in Figure 6.2.

3. Decrypt data and authenticate the message.

   The encrypted message-receiving entity, whether AP or adapter, performs the reverse steps to recover the original data and authenticate that the message was sent by someone with whom the recipient has a shared key. In decryption, the IV from the incoming message along with the shared key (remember we are using a shared key and both peers are using identical shared keys) is used to generate the encryption key (as in Step 1), which is then used to decrypt the incoming message by XORing the encrypted message with the encryption key. The steps are as shown in Figure 6.3.

4. Authenticate the received message.

   Performing the integrity check algorithm on the recovered plaintext and comparing the output CRC32 algorithm with the last 32 bits of the transmitted data verifies the decryption and authentication. If the calculated CRC is not equal to the CRC value received in the message, the received message is in error, and an error indication is sent to the media access control (MAC) management and back to the sending station (see Chapter 2 for more information on MAC protocol). Mobile units with erroneous messages (due to inability to decrypt) are not authenticated.

   ```
   CRC-Value = CRC32(Decrypted-Message)
   ```

   The same shared key used to encrypt/decrypt the data frames is also used to authenticate the station. It is considered a security risk to have both the encryption keys and authentication keys be the same. There is also a method where users and APs can utilize WEP alone without shared-key authentication, essentially using WEP as an encryption engine only. This is done in open system mode. This is considered to be the most protected implementation in 802.11 thus far and still enables reasonable authentication.

```
MessageReceived = POLOJMNB

EncryptedMessage = MessageReceived - First-24-bits

IV = MessageReceived - Last-40-bits

Concatenated-Key = Shared-Key + IV

Decryption-Key = RC4(Concatenated-Key)

Decrypted-Message = (EncryptedMessage) XOR Decryption-Key
```

Figure 6.3 Data encryption using WEP.

802.11 WEP Authentication

WEP provides two authentication modes: open-system authentication and shared-key authentication.

Open-System Authentication

The open-system authentication is also known as null authentication because a wireless LAN adapter can associate with any access point and listen to all the data that is sent in plaintext. This is usually implemented where ease of use is the main issue, and the network administrator does not want to deal with security at all. This is the default authentication service that does not have authentication.

Shared-Key Authentication

This involves a shared secret key to authenticate the wireless LAN adapter to the AP. The shared-key authentication approach provides a better degree of authentication than the open system approach. For a station to utilize shared-key authentication, it must implement WEP Encryption Protocol, as discussed earlier. Figure 6.4 illustrates the operation of shared-key authentication. The 802.11 standard does not specify how to distribute the keys to each station, however. The process is as follows:

1. A requesting wireless LAN adapter sends an authentication frame (a frame is a data of fixed length) to an AP it wants to authenticate.
2. When the AP receives an initial authentication frame, the AP will reply with an authentication frame containing 128 bytes of random challenge text generated by the WEP engine in standard form.
3. The requesting wireless LAN adapter will then copy the challenge text into an authentication frame, encrypt it with a shared key, and send the frame to the responding station.
4. The receiving AP will decrypt the value of the challenge text using the same shared key and compare it to the challenge text sent earlier.
5. If a match occurs, the responding wireless LAN adapter will reply with an authentication indicating a successful authentication. If not, the responding AP will send a negative authentication.

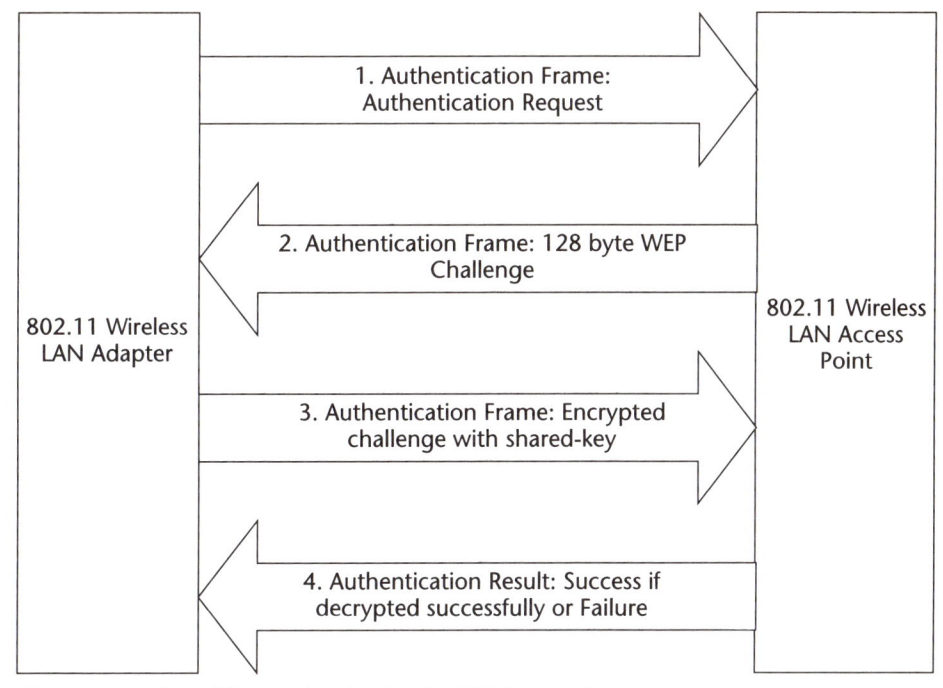

Figure 6.4 Shared-key authentication in WEP Protocol.

IEEE 802.11 WEP Protocol Weaknesses and Shortcomings

WEP can be easily cracked in both 40- and 128-bit variants by using off-the-shelf tools readily available on the Internet. As of the time this book was written, on a busy network, 128-bit static WEP keys can be obtained in as little as 15 minutes. Besides the shared-key weakness that WEP suffers, some of the other known vulnerabilities of WEP are as follows:

No per-packet authentication. Subsequent frames transmitted after the authentication frame do not contain any authentication data.

Vulnerability to disassociation attacks. Disassociation is where a wireless LAN adapter terminates its communication with an AP. In disassociation attacks, an adversary injects forged packets into a wireless LAN, requesting that a valid wireless LAN adapter be disassociated, effectively requiring the valid adapter and AP to perform reauthentication.

No user identification and authentication. The authentication and identification supported in the 802.11 standard provide only MAC-level authentication and identification. The actual user of the network device is never authenticated.

No central authentication, authorization, and accounting support. Each AP manages its own authentication, authorization, and accounting (logging activities). If more than one AP is used, the effort involved in managing APs is a factor of the number of APs used.

RC4 stream cipher is vulnerable to known plaintext attacks. RC4 is considered unsafe due to known cryptographic attacks. Though these attacks require a significant amount of processing power, the insecurity of using RC4 adds to the vulnerability of the WEP protocol.

The initialization vectors (IVs) are at the center of most of the issues that involve WEP. Because the IV is transmitted as plaintext and placed in the 802.11 header, anyone sniffing a WLAN can see it. At 24 bits long, the IV provides a range of 16,777,216 possible values. A University of California at Berkeley paper found that when the same IV is used with the same key on an encrypted packet, known as an IV collision, a hacker could capture the data frames and derive information about the data as well as the network. For more information, refer to the paper at: http://www.isaac.cs.berkeley.edu/isaac/wep-faq.html.

In addition to the weaknesses found in the WEP protocol by the University of California at Berkeley, recently cryptanalysts Fluhrer, Mantin, and Shamir discovered inherent shortcomings with the RC4 key-scheduling algorithm [5]. Because RC4 as implemented in WEP chose to use a 24-bit IV and does not dynamically rotate encryption keys, these shortcomings are demonstrated to have practical applications in decrypting 802.11 frames using WEP. The attack illustrated in the paper focuses on a large class of weak IVs that can be generated by RC4, and highlights methods to break the key using certain patterns in the IVs.

The WEP protocol is, therefore, considered insecure due to the improper use of initialization vectors and the key scheduling as defined in the WEP protocol, and the lack of authentication primitives for both packet and user-based authentication.

The Future of 802.11 Standard Security

IEEE 802.11 is currently working on extensions to WEP for incorporation within a future version of the standard. This work was initiated in July 1999 as Task Group E, with the specific goal of strengthening the security mechanisms so as to provide a level of security beyond the initial requirements for WEP. The enhancements currently proposed are intended to counter extremely sophisticated attacks, including those that have been recently reported in the press. In addition it needs to be noted that the choice of encryption algorithms by IEEE 802.11 are not purely technical decisions, they are limited by government export law restrictions as well.

Common Security Oversights

Most wireless LAN equipment is shipped with security features disabled. To enable security on 802.11 devices, care must be taken to ensure proper security of the wireless LANs. In this section, we examine the most common security oversights that degrade the security of a wireless LAN.

Using Default or Out-of-the-Box Security

As mentioned earlier, most currently available wireless LAN devices, especially 802.11-compliant devices, come with security features disabled, and under most circumstances the wireless LANs are deployed without enabling the security features. This widespread deployment of insecure wireless LANs has attracted the attention of the hacker community. An unprotected network may also provide outsiders free access to its broadband access. There is a parasitic activity commonly referred to as war driving that hackers engage in, where the primary purpose is to use the Internet services of other individuals and corporations. War driving is an adaptation of another activity known as war dialing. War dialers use brute force to dial every phone number looking for modems, trying to break into systems and network. A war driver generally roams neighborhoods, office parks, and industrial areas looking for unprotected networks and sometimes sharing this information on the Internet. To protect a wireless LAN from hackers and other adversaries, it should always be operated in encrypted and authenticated mode.

Using Fixed Shared Keys

Most currently available wireless LAN devices support more than one shared key. These shared keys are used for authentication and encryption purposes. If these keys are not frequently updated, they might be hacked. Therefore, to ensure that a wireless LAN is secured, the shared keys should be updated on a frequent basis to avoid shared-key–based attacks.

Using Far-Too-Strong Radio Signals

The strength of radio signals used in a wireless LAN define the range from which a wireless LAN can be accessed. Use of devices that produce stronger signals than are needed add insecurity to a wireless LAN as they become accessible to adversaries from farther distances. It is, therefore, important to use wireless devices that emit radio signals that are not too strong.

Extending Wireless LAN Security

In deployment scenarios where securing wireless LANs is crucial and the 802.11-standard wireless LAN security does not seem enough, alternate security measures can be adopted to provide a higher level of security. These measures include authentication and privacy mechanism at network level by using supplementary technologies like 802.1X and virtual private networks (VPNs). In this section, we look at the ways in which the 802.1X authentication protocol and the VPN can be used to improve the security provided by the 802.11 standard.

The 802.1X Authentication Protocol

The 802.1X is an IEEE draft stand that defines a port-based network access control protocol (that is, one involving a network that uses more than one channel to perform network operations instead of using one channel for all operations). 802.1X was originally designed for Ethernet-based LANs, but it can also be applied to 802.11-based wireless networks. 802.11 does not require that all LAN devices use the same WEP keys, and allows a device to maintain two sets of shared keys: a per-station unicast session key and a multicast/global key. Current 802.11 implementations primarily support shared multicast/global keys, but are expected to support per-station unicast session keys in the near future. Managing and updating all of these keys can be a difficult manual process, and it does not scale appropriately in large infrastructure network or in an ad-hoc network. In addition, the lack of an interaccess point protocol (IAPP), a protocol that will facilitate communication between two APs, further compounds key management issues when wireless

LAN devices roam from one AP to another, since without this protocol, authentication has to begin anew.

The Basic 802.1X Operation

To understand the basic operation of 802.1X, let's define some of the entities involved in an 802.1X authentication protocol. These entities are authenticator, supplicant, and the authentication server. An authenticator is an entity that enforces authentication before allowing access to services. The supplicant is an entity that requests access to services available via the authenticator. An authentication server performs an authentication function: It checks the credentials of the supplicant on behalf of the authenticator. The authentication server then responds to the authenticator indicating whether or not the supplicant is authorized to access the authenticator's services. The authentication server may be a separate entity, or its functions may be colocated with the authenticator. The most widely used authentication server is the Remote Authentication Dial-in User Service (RADIUS) server. Figure 6.5 shows a basic arrangement of 802.1X entities.

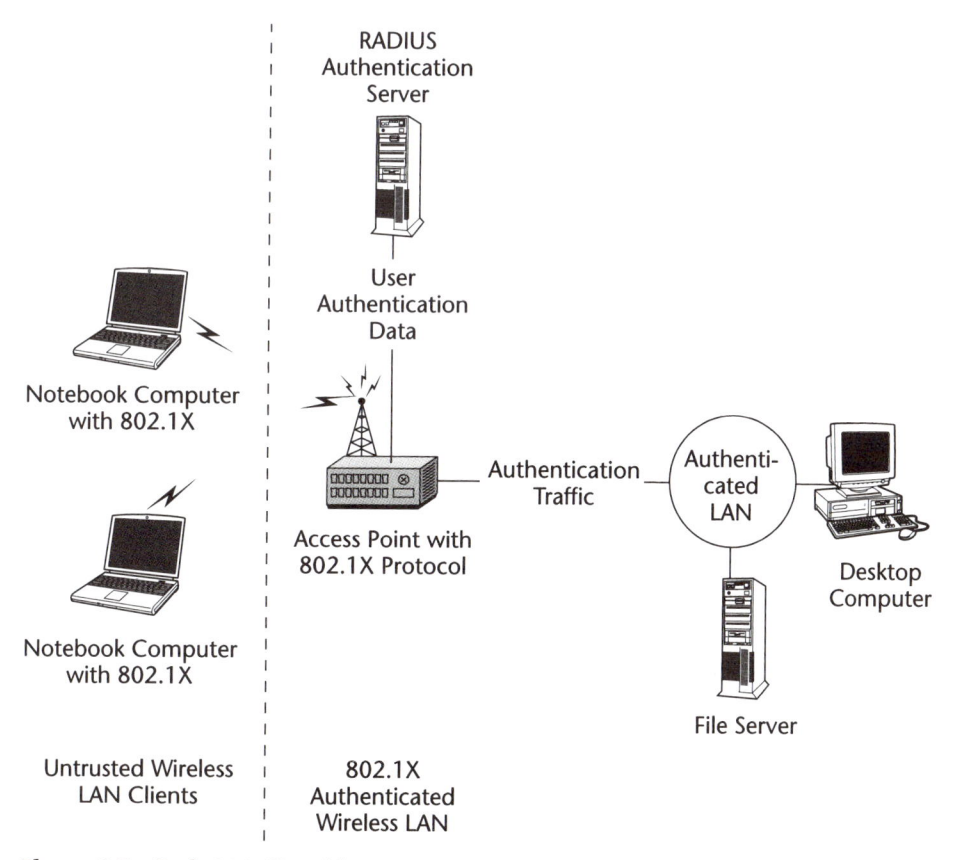

Figure 6.5 Basic 802.1X entities.

A LAN port can play one of two roles in a network access control interaction: authenticator or supplicant. The authenticator's port-based access control defines two logical access points to the LAN via a single, physical LAN. The first logical access point, labeled Uncontrolled Port, allows an uncontrolled exchange between the authenticator and other systems on the LAN—regardless of the system's authorization state. The second logical access point, labeled Controlled Port, allows an exchange between a system on the LAN and the authenticator's services—only if the system is authorized.

One use of the uncontrolled port would be to provide a path for exchanges between the authenticator and the supplicant. The authorization state of the controlled port determines whether traffic can flow from the supplicant to the LAN through this port. The authorization state will likely start as unauthorized, and then transition to the authorized state upon authentication of the supplicant.

802.1X typically uses Extensible Authentication Protocol (EAP) as a means to communicate the authentication information between the supplicant and authentication server. This means that EAP messages need to be encapsulated directly over a LAN medium. Another protocol, EAP over LAN (EAPOL), was defined for this purpose.

An Example 802.1X Exchange for Authentication

In this example, we assume the use of EAP as the protocol for exchanging the authentication data, and a RADIUS server for an authentication server. There are several other possible message flows depending on the authentication mechanism used. An example exchange that could take place to authenticate the supplicant might be as follows:

1. The authenticator sends an EAP–Request/Identity message to the supplicant.
2. The supplicant sends an EAP–Response/Identity with its identity to the authenticator. The authenticator forwards this to the authentication server.
3. The authentication server responds with an EAP–Request packet containing a password challenge to the supplicant through the authenticator.
4. The supplicant sends its response to the challenge to the authentication server through the authenticator.
5. If the authorization is successful, the authorization server sends an EAP–Success response to the supplicant through the authenticator. The authenticator can use this success to set the controlled port state to authorize.

Using 802.1X to Solve the 802.11 WEP Security Issues

The basic 802.1X protocol must be extended to address security issues of 802.11. This is done by passing an authentication key to the client, and to the wireless access point, as part of the authentication procedure. Only an authenticated client knows the authentication key, and the authentication key encrypts all packets sent by a client as defined in the WEP protocol. 802.1X helps the WEP problems by providing keys per station or per session to limit the number of packets using the same key, and by making sure that the keys are changed often—rekeying (changing the keys) as much as every 5 to 10 minutes or 4 million packets, therefore limiting the reuse of shared keys, the main weakness of the WEP protocol. With 802.1X implemented and deployed, this can be accomplished automatically. Following authentication, the 802.1X protocol should be configured to request that the station reauthenticate periodically, at a specific time interval. 802.1X thus provides per-station, per-session keys, and causes these keys to be changed often, eliminating reuse issues. In addition, 802.1X allows for user identification and authentication and centralized authentication, authorization, and accounting support. This also allows for the future use of extended authentication mechanisms.

It is important to remember that all authentication traffic is communicated through the uncontrolled port, whereas all authorized data transfer takes place on the control port once the user has been authenticated. Authentication of a wireless LAN adapter using 802.1X protocol consists of the following steps:

1. Without a valid authentication key, an AP inhibits all traffic flow through it.

2. When a wireless LAN adapter (supplicant) comes in range of a wireless AP authenticator, the wireless AP issues a challenge to the wireless station.

3. Upon receiving the challenge from the AP, the wireless LAN adapter responds with its identity.

4. The AP then forwards the wireless LAN adapter's identity on to the RADIUS server to initiate authentication services.

5. The RADIUS server then requests the credentials for the station, specifying the type of credentials required to confirm the wireless LAN adapter's identity.

6. The wireless LAN adapter sends its credentials to the RADIUS server.

7. Upon validating the wireless LAN adapter's credentials, the RADIUS server transmits an authentication key to the AP. The authentication key is encrypted so that only the AP can access it.

8. The AP uses the authentication key received from the RADIUS server to securely transmit—per-wireless LAN adapter unicast session and multicast/global authentication keys—to the station. This key is always transmitted in encrypted format.

Virtual Private Networks (VPNs)

Virtual private networks (VPNs) are typically used in TCP/IP-based networks to secure communication between remote users and a private network. A typical usage scenario for a VPN can be a remote worker who uses a dialup connection from his or her home to connect to the Internet and uses VPN to establish a secure network session with the corporate network at the company he or she works at. Using VPN to establish such connectivity guarantees that the remote user is authenticated and all data over the Internet is transmitted in encrypted form. The usage scenario we just discussed is shown in Figure 6.6.

VPN technology can be used in the wireless LANs to provide user authentication and data privacy just like a remote user that accesses a corporate LAN over the Internet using VPN. Some confusion may arise in deciding when to use a VPN and when to use 802.1X. These technologies actually complement each other, and there are times when both may be used. VPN can be used in a variety of ways with 802.11-based wireless LANs to provide security between a user and the extended wired LAN, among users in a wireless LAN, and between a wireless user and an AP connected to a corporate LAN over the Internet. The combinations are shown in Figure 6.7.

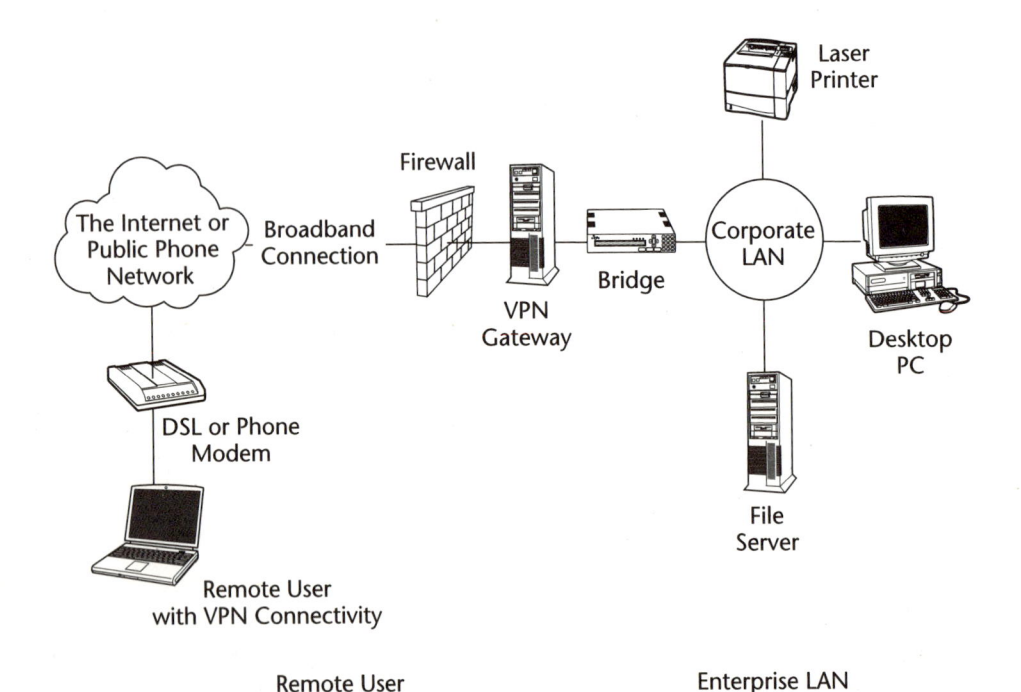

Figure 6.6 A remote user connected to corporate LAN over the Internet using VPN.

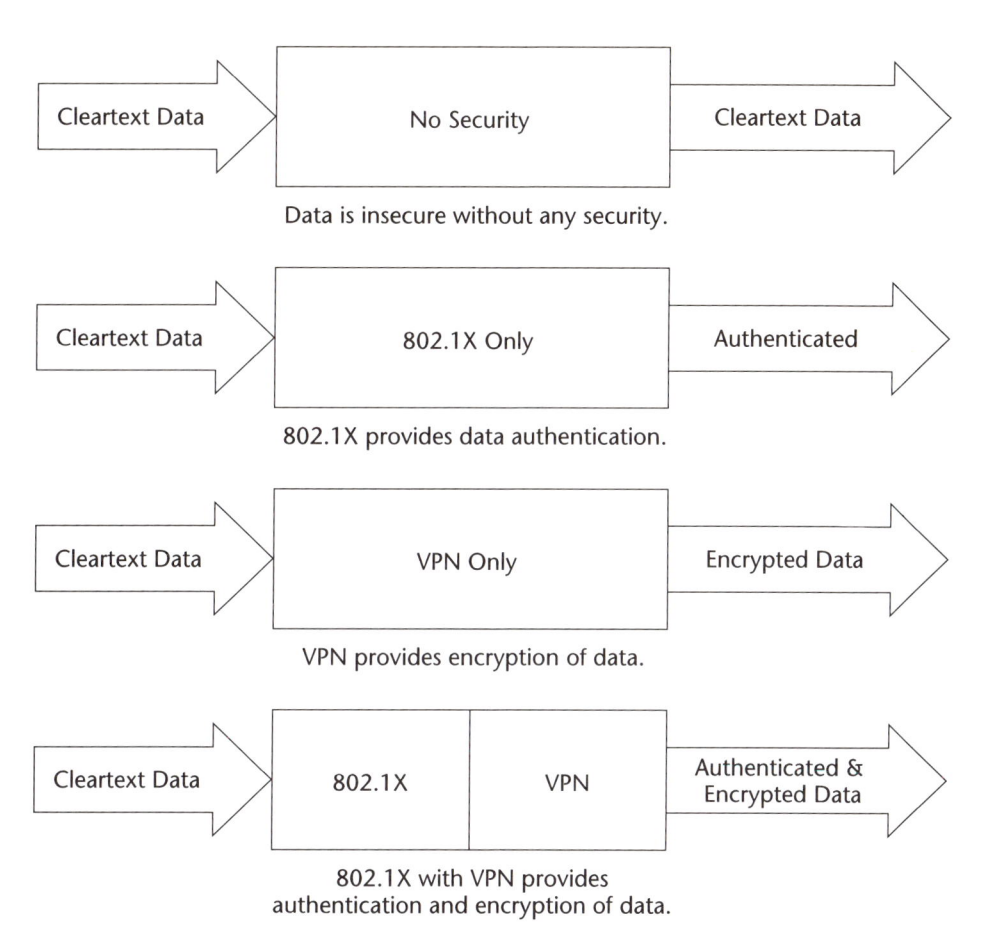

Figure 6.7 Various combinations of securing data over wireless LANs.

Securing Wireless LAN

The primary and foremost fact to remember when securing an 802.11 wireless LAN is that 802.11 devices are shipped with all security features disabled. It is the responsibility of those involved in deployment to ensure that appropriate security measures are taken. A secured wireless LAN includes provisions for authentication of devices and users, all APs are managed by using proper security, and the data is transmitted in an encrypted form. In this section, we talk about the measures that are necessary to ensure proper security of a wireless LAN.

User Authentication

All users in a secure wireless LAN must be authenticated. They can be authenticated using WEP-based authentication for minimal authentication security, 802.1X for moderate security, and VPNs for high-level security.

In addition to the wireless LAN authentication security, the LAN devices using the wireless LAN, or any other networks that wireless LANs may have access to, must enforce operating system (OS)–level security requiring at least username and password to authenticate the users in the LAN. The OS-level authentication is necessary because it supplies security to the network resources that might be available in a network. For example, a file server on a LAN must require OS authentication to allow only those users access to the data on the file server that have been successfully authorized by the OS authentication mechanism to ensure that the users attempting to access the files are in fact authorized to do so.

802.1X can be used for authenticating wireless LAN users on a given wireless LAN. Mostly 802.1X will be used to authenticate users in a LAN environment in which a wireless LAN is connected with a wired LAN, and wired LAN provides authentication services to the APs connected to it.

VPNs must be deployed any time data security is critical, especially when a wireless LAN device connects to a remote LAN using an AP connected to the Internet. In this scenario, an AP must not be connected with the remote LAN directly; instead the users of the wireless LANs who establish direct connection between their devices and the remote corporate LAN must use VPN to establish such connections.

VPNs and 802.1X can be combined to provide wireless and wired LAN security to networks that include both local LANs and remote LANs. In this case all wireless devices are authenticated using the 802.1X protocol, and VPN is used to provide enhanced data security between the wireless devices, local wired LANs, and the remote LANs.

Data Confidentiality and Privacy

For a wireless LAN to be called a secured LAN, all traffic through the LAN must be properly transmitted in an encrypted form. The data between wireless devices and the AP can be secured using WEP security for minimal security, and using VPN technology to provide high-level security.

WEP protocol uses the RC4 algorithm to provide data confidentiality and privacy (see *The Institute of Electrical and Electronics Engineers (IEEE) 802.11 Standard Security* earlier in this chapter). The security experts have heavily criticized the insecure use of RC4 in WEP protocol, where the initialization vector and encryption keys are considered the weakness of the protocol. 802.1X solves the problems in the basic WEP protocol by providing a better mechanism for changing keys and authenticating users. For minimal security needs, the basic WEP encryption may be used with extreme caution, but if more reliable and medium-level security is desired, the 802.1X-based security primitive must be used.

For high-level data confidentiality and privacy, VPNs must be used. VPNs provide data confidentiality by encrypting all data that is transmitted by communicating entities. VPNs can also be used with 802.1X to restrict only authorized users' access to the wireless LANs, thereby allowing access to LAN resources to only those users who are authorized by 802.1X security and the VPN security provisions.

Wireless LAN Passwords and Usage Policies

Users and administrators of any LAN, especially wireless LANs, must be required to regularly change passwords. All users should be encouraged to use passwords that are hard to guess. Users must be strongly discouraged from sharing their passwords with other individuals. Access to resources requiring high security must be restricted, and users must not be allowed to use wireless LANs with security features turned off.

Frequent Network Traffic and Usage Analysis

Administrators of wireless LANs should monitor network traffic and usage on a regular basis to ensure that network security is not compromised. Authentication logs must be frequently observed to identify any security breaches or any attempts that are targeted to attack a network.

Summary

Today wireless LANs are rapidly being deployed without much regard to the security risks that they introduce. The currently available 802.11-standard devices include WEP protocol–based security, which is almost always disabled when a manufacturer ships the device. The situation is further weakened by the known weaknesses in the WEP protocol. Hence, besides enabling the security features of WEP, depending on the security requirements, alternate protocols and extensions to the WEP protocols must be explored to provide the desired level of security when building a secure wireless LAN. Common ways to extend a wireless LAN include the enforcement of security policies and the use of the 802.1X protocol and/or VPNs to provide authentication and data security.

In the next chapter, we help you plan a wireless LAN deployment. We help you identify your needs, performing a site survey, setting up reasonable expectations for wireless LANs, and estimating the basic hardware and software that you may need. We also walk you through an example of wireless LAN planning that will help you understand the planning process for a wireless LAN.

PART Three

Building Secure Wireless LANs

Building a secure wireless LAN is a challenging task. It requires a very good understanding of basic computer networking concepts, wireless LAN technologies and standards, and the specific security requirements of wireless LANs. This book is written to guide an individual through the steps necessary to build a successful secure wireless LAN based on the IEEE 802.11 standard. In Part 1 of the book we described basic wired LANs to introduce the networking concepts that are necessary to understand wireless LAN. We introduced the wireless LAN technology, briefly discussed the IEEE 802.11 standard, and talked about the strengths and the weaknesses of wireless LANs to help you decide whether wireless LANs are right for you. After reading Part 1 you should have a good understanding of wired LANs, wireless LANs, the current 802.11 standards, and whether or not wireless LANs are right for you. In Part 2, we first examined the security issues of wired LANs. Then we discussed the security issues and risks surrounding wireless LANs and the primitives that can be used to assure a higher level of security than that defined by the 802.11 standard.

Part 3 of the book utilizes the knowledge that we built in Parts 1 and 2 and guides you through building secure wireless LANs using the 802.11 standard. We have divided the process of building a wireless LAN into three separate steps: planning a wireless LAN, purchasing the right equipment, and setting up the equipment. We also discuss the currently available means to securely connect a wireless LAN to an enterprise network.

Chapter 7 talks about the significance of planning a wireless LAN. We help you make the basic decisions that help you build an extensible and flexible wireless LAN.

Chapter 8 helps you decide what kind of wireless LAN equipment you will need for a particular deployment scenario. We talk about equipment selection based on SoHo, Enterprise, and WISP scenarios.

Chapter 9 discusses the actual process of setting up wireless LANs. In this chapter we help you design a wireless LAN that provides a secure operation and suits your needs.

Chapter 10 explains how to extend a wireless LAN by connecting it with an enterprise LAN using a virtual private network.

When you finish reading Part 3, you will be able to successfully build a wireless LAN. You will understand the process of planning a secure wireless LAN, and the equipment that fits a particular deployment scenario. You will also be able to set up a wireless LAN, and interconnect a wireless LAN to an enterprise LAN using VPN.

CHAPTER 7

Planning Wireless LANs

The best thing about wireless LANs is that they are extremely easy to build compared to their wired counterparts. However, wireless LANs require far better understanding of the technologies involved and careful planning to provide a flawless, high-performance, and secure operation. An unplanned wireless LAN could result in severe problems due to bad performance and security issues. In order to plan a secure wireless LAN, you must have a good understanding of the following:

Basic networking concepts. Except for the physical transmission characteristics of wireless LANs, wireless LANs are almost identical to wired LANs. All the basic networking concepts that apply to one also apply to the other. Therefore, it is important to understand basic networking concepts. If you are new to computer networks, we suggest that you read Chapter 1 of this book, which explains basic networking concepts.

Wireless LAN technologies, basic equipment, and standards. Wireless LANs are based on relatively new technologies when compared to wired LANs, which have been around for a while and are well understood. The technologies involved in wireless LANs are constantly evolving, resulting in newer equipment and standards. When selecting wireless LAN equipment, it is important that all available technologies are evaluated to ensure the best possible performing LAN that fits your needs and budget. Chapters 2 and 3 talk about the basic wireless LAN technologies and standards that exist today.

Security issues with wireless LANs. When deploying a secure wireless LAN, security primitives of the technologies must be well understood. Chapters 5 and 6 detail the security concerns of wireless LANs.

Whether wireless LANs are right for you. You might be able to answer this question by reading previous chapters, or you may be able to better answer this question when you have finished planning your wireless LAN.

Documenting the planning stages is one of the most important tasks when planning a LAN deployment. Documenting the planning steps ensures that you follow a guideline. Unfortunately, this step is often skipped when networks are planned, resulting in the realization, perhaps too late, that key factors or important tasks and objectives have been overlooked. We suggest that you take notes and document your planning process as you follow each step in this chapter.

Let's start planning for a wireless LAN. There are seven basic steps involved in planning a secure wireless LAN.

Step 1: Understanding Your Wireless LAN Needs

The primary and foremost step when planning a wireless LAN is to understand your own wireless LAN needs. This ensures that you have accounted for most LAN services that you desire to use in a wireless LAN; it also helps you select the type of wireless LAN that satisfies your needs. In Step 1, we help you identify the services you want from a wireless LAN, and help you plan for the type of wireless LAN that might best fit your needs.

Identifying the computer programs and applications that you will be running on the computers accessing the wireless LAN is of critical importance as it helps you decide whether wireless LANs satisfy your primary needs or not. Understanding your LAN requirements is of even more importance where a wireless LAN is aimed to replace an existing wired LAN. In this scenario, if the desired services are not considered in the planning phase, the resulting wireless LAN could lack the basic features that were available in the wired LAN. We suggest that you inventory and identify all the software applications that need to access data and resources on the network. Special attention should be given to applications that use streaming audio and video as well as client applications accessing large amounts of data.

File and Application Servers

Most LANs are used for sharing files among users. These files reside on the large-capacity hard disks on computers, which can be standalone file servers or workstations, and are connected to the network. In addition to file-sharing services, network-based application programs must be considered. These application programs could be email servers, accounting software, or a Web server.

Considering file and application servers is important because if wireless LAN hardware or software will not properly support the computers that provide these services, the entire wireless LAN could easily become useless.

Print Servers

The ability to share printers is another big use of LANs. If a printing feature is to be supported on the wireless LAN, ensure that it is accounted for when planning for wireless LANs. When setting up the wireless LAN, you will have to decide how to connect printers to the wireless LAN.

The Internet Access

Internet access has become one of the prime uses of LANs. LANs normally share a high-speed connection among users. You should consider how you want to share the Internet connection among users of a wireless LAN. Depending on the deployment scenario, desired connection bandwidth must be considered in order to provide a high-bandwidth solution to the users of wireless LANs.

Miscellaneous Network Appliances

All other devices that are to be attached to a network must be considered when defining the services you need from a wireless LAN. Such devices could include a modem pool, tape backup, or site-security systems.

Select Suitable Type of Wireless LAN

Selecting the type of wireless LAN you need depends on a number of factors, including: the number of intended wireless LAN connections, the deployment scenario (for example, home use, SoHo, enterprise, or WISP), security, and the external connectivity desired. Here we discuss some of the basic types of wireless LANs that you should consider when planning a wireless LAN.

Peer-to-Peer or Ad-Hoc Wireless LAN

Peer-to-peer wireless LANs consist of two or more computers, generally without any type of server (for example, file server or mail server). Instead the computers communicate directly with each other. Such wireless LANs have limited use and are only deployed in small LAN settings like home or SoHo environments.

Standalone Wireless LAN

Standalone wireless LANs consist of one or more computers and may include a file server and other shared network resources. These LANs are considered relatively secure as they are normally not connected to the Internet or any external network. Such networks are generally used in SoHo, retail businesses, and manufacturing environments.

Wireless LAN as a Replacement for a Wired LAN

Many organizations are replacing their wired LANs with wireless LANs. Creating a wireless LAN in place of a wired LAN provides mobility to the users and makes future LAN upgrades a little easier. For example, to upgrade a wired LAN, you often have to replace the physical wiring, whereas wireless LANs do not have this requirement. To upgrade wireless LAN technology, only wireless LAN equipment and software need to be upgraded and the new LAN becomes operational. Total replacement of a wired LAN with a wireless LAN is a big task and should be carried out with great caution. Conduct a small pilot first before rolling out the full-blown wireless LAN.

Wireless LAN as an Extension to a Wired LAN

A good idea when deploying a wireless LAN is to first deploy it as an extension to an existing wired LAN. This means that you leave the wired LAN intact and add a wireless LAN as a supplementary network to the original LAN. This allows a smooth division of the existing LAN, in which the computer systems that are required to be on the wired LAN stay on the wired LAN and those needing wireless connectivity are upgraded with the wireless LAN equipment.

It is, however, important to remember that the wireless LAN technology available today operates at slower speeds than its counterparts, the wired LANs. Therefore, it is often a good idea to deploy a wireless LAN as an extension to a wired LAN, where all computers and devices requiring high network bandwidth are connected to the wired LAN and the workstations or peripherals that require lower bandwidth are connected using the wireless equipment.

Once you understand your needs for wireless LANs, you are ready to plan the scope of rollout and set up the requirements and expectations for the LAN you are willing to deploy.

Step 2: Planning the Scope of Rollout

As the second step, you should define the rollout scope of the wireless LAN you are planning to deploy. You should decide whether you want to deploy the entire wireless LAN at once or begin by first doing a pilot and then progressively roll out the actual wireless LAN. The incremental rollouts help you better understand your needs and allow you to choose the right technologies.

We suggest that you first plan a pilot then roll out a full-blown wireless LAN. If you are replacing a wired LAN with a wireless LAN, you might want to try first using wireless LAN in conjunction with the wired LAN and then replace the actual wired LAN to avoid any negative surprises.

Step 3: Performing Site Survey

For a comprehensive plan, you must perform a site survey for each physical location where you intend to install a wireless LAN. The site survey includes careful consideration of the geographic coverage area, per-site security requirements, and profiling wireless LAN users and devices.

Considering the Geographic Coverage Area

Wireless LAN signals are susceptible to interference from other competing devices that use the same frequency bandwidth. For example, 802.11b operates in 2.4-GHz ISM band and is vulnerable to interference from microwave ovens, cordless phones, and Bluetooth-based devices. In addition to the interference-related problems that wireless LANs suffer, steel objects and thick walls easily obstruct wireless LAN signals. The interference and obstruction in the wireless LAN data path reduce the performance of a wireless LAN. Performance degradation also occurs when the distance between a wireless LAN device and an access point (AP) increases. Such problems can be answered by carefully locating the best spots for AP placement, locating dead spots (areas where wireless LAN signals cannot reach), and ensuring that the least number of competing devices operate in the region where you are planning to roll out a wireless LAN.

Locating the Best Spots for Wireless Access Points

The best spots for installing wireless LAN access points are the areas that allow the least obstructed signal transmission and are the closest to the wireless LAN users. To locate the best spots, the signal strength should be monitored using a signal monitoring software. Most wireless LAN equipment comes with signal monitoring software that can show you the relative signals at various distances from a given AP. You should plan on examining signals from each AP that you intend to install at a given site. Plan to install APs at distances such that the APs barely overlap the coverage area. Installing APs this way ensures that a mobile user roaming between two APs always receives a wireless LAN signal.

Locating Dead Spots

Dead spots in a wireless LAN are physical regions where wireless LAN signals cannot reach due to the nature of physical construction (for example, steel locker or a bunker-like construction) or excessive interference. All such dead spots must be identified, and a decision should be made whether it is worth providing wireless services to these areas. If wireless services are necessary for those areas, either additional APs should be installed or a wireless LAN technology should be selected that is less vulnerable than originally planned. If the availability of wireless LANs is not necessary in the dead spots, you should consider using wired LAN extensions for LAN connectivity at the dead spots.

Per-Site Security Requirements

If a wireless LAN is quite large, consists of one or more floors, or is located in an area that requires high security, a blanket security policy might not be the best approach. Different regions on a wireless LAN may be planned to use different security primitives, where some will require high security and others lower security. For example, regions on the outer periphery of a site may require longer encryption keys, whereas inner regions may require shorter encryption keys.

Profiling Wireless LAN Users and Devices

If all LAN devices and users are to be connected using wireless LAN equipment, for instance in the case of a standalone wireless LAN, then this step may be skipped. In all other cases a thorough and careful examination of all computers and devices that will operate in the wireless LAN must be performed to ensure better planning for the equipment that wireless LAN users might need. You should profile users and devices based on the following criteria:

Nature of use. Users of wireless LANs who travel with their computers and have mobile computing requirements are better candidates for wireless LANs, whereas users who only work at their desks are not the best candidates for wireless LANs.

Computing device type. Users with mobile computing devices (for example, laptops, notebooks, and handheld personal digital assistants [PDAs]) are true users of wireless LANs, whereas users with desktop computers and workstations may not need wireless LAN capability. However, in circumstances where wireless LANs are being deployed to replace wired LANs, this qualification may not apply and all computers may be planned to use wireless LAN.

Operating system (OS). Generally speaking, all computing devices are installed with an operating system (OS). All such devices should be carefully accounted for, as all OSs might not support the wireless LAN technology that you are planning to use. If a scenario is noticed where an OS does not support the intended wireless LAN technology, you should contact the OS vendor or plan to install a different operating system on such devices. For example, if you encounter a computer system with Windows 3.1 installed and you are using 802.11b wireless LAN technology, it is most likely that the drivers for the 802.11b might not be available for Windows 3.1 and you should plan on upgrading the operating system or use wired LANs to connect this computer to the LAN.

Bandwidth requirements of the computers. If computers that need to be in wireless LANs require high network throughput (for example, a file server), you might try using a wireless LAN technology that allows such throughput. For example, 802.11b supports speeds up to 11 Mbps, but if you have higher speed requirements, you might want to think about using 802.11a-standard LAN equipment that provides operating speeds of up to 54 Mbps.

Step 4: Setting Up Requirements and Expectations

When building a wireless LAN, it is important to establish practical expectations while setting requirements for the LAN. When setting requirements, you should make sure that they are practical and possible given the technologies you are planning to use. For example, you cannot expect 1,000 users to simultaneously use an AP and maintain the 11-Mbps speed when using an 802.11b wireless

LAN, as most 802.11b APs support a maximum of 64 users. However, you should establish the minimum requirements for the LAN. This will help you carefully choose equipment and set up your expectations.

Network Bandwidth and Speed

Because wireless is a shared medium among all wireless LAN devices on a LAN, the network bandwidth and speed degrade with any increase in the number of users that share an AP. When planning a wireless LAN, you must consider whether the throughput of the proposed wireless LAN technology is sufficient for the number of users. For example, as explained in the last section, the 802.11b-based wireless LANs operate at 11 Mbps, and if the users need a bandwidth of 100 Mbps, 802.11b might not be the right choice. When planning, therefore, you must compare the actual needs of users with the technologies available and carefully communicate the characteristics of the planned network with those who will be using the wireless LAN.

Coverage Area and Range of Wireless LANs

Though many wireless LAN equipment manufacturers claim that their devices can operate at ranges up to 300 meters, more often a range like this is only possible in open ground under conditions of direct visibility. Steel structures and thick walls easily obstruct wireless LAN signals, traveling over the airwaves. Wireless LANs operate better in environments that lack obstructions. For example, wireless LANs in a building with cubicles will perform better than in a building with individual offices. In addition, if a location where wireless LANs are to be deployed consists of many floors, the signals may be obstructed by the ceilings and the floors of the building.

Security

The current security standards of wireless LANs are highly criticized. While deploying wireless LANs, you should thoroughly understand the weaknesses and strengths of wireless LAN security to avoid making claims that might not turn out to be true in the long term. Be advised that current wireless LAN security standards are vulnerable to attacks. For more information on wireless LAN security issues, see Part 2 of this book, "Secure Wireless LANs."

Step 5: Estimating the Required Wireless LAN Hardware and Software

In this section, we talk about estimating the required hardware and software that you will need to deploy your wireless LAN successfully.

Basic Wireless LAN Hardware

The wireless LAN hardware that you might need will depend on your deployment scenario. You should plan on using the hardware that suits your requirements, is right for your budget, and is upgradable in future. In this section, we talk about some of the most commonly used LAN equipment in a wireless LAN deployment.

Access Points (APs)

Access points are considered to be the heart of wireless LANs as they facilitate the data transmission among computers in a network. (For more information on APs, see Chapter 2, "Wireless LANs.") The AP devices are wireless-LAN-technology specific, and an AP based on one technology normally does not operate with another technology. For example, an 802.11b wireless LAN adapter does not work with an 802.11a-based AP. It is important to realize that a wireless LAN that consists of 802.11b is a totally separate LAN from one constructed using the 802.11a technology, even if they share exactly the same physical area.

A carefully conducted site survey should help you plan the number of APs that are needed at a site. If your site survey suggests that you need a hybrid wireless LAN where both 802.11b and 802.11a devices must coexist, you should plan to connect the APs forming two distinct LANs and interconnect the two LANs using a wired LAN technology (see Figure 7.1). This will ensure that even though you have built a hybrid LAN, users on either LAN can communicate with each other.

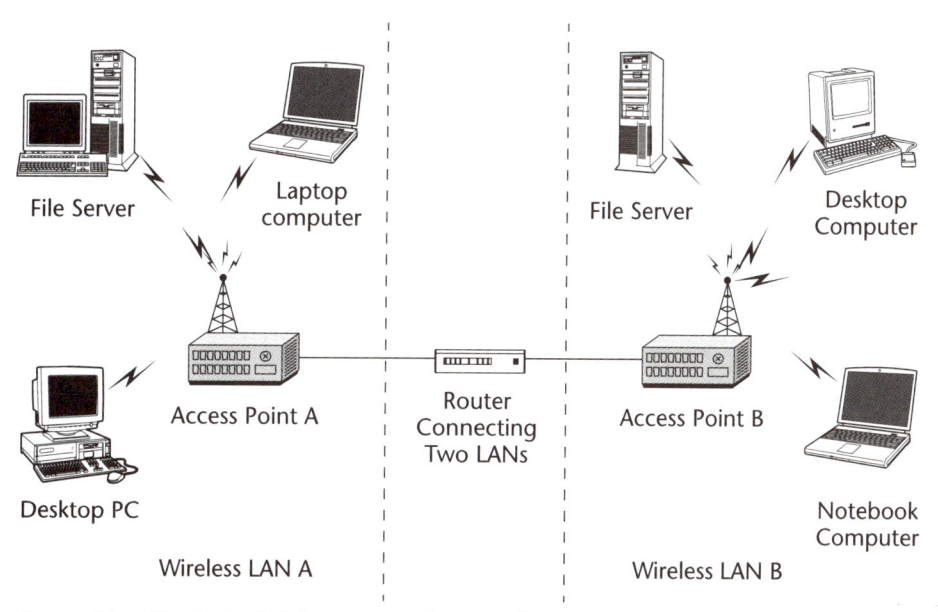

Figure 7.1 Wired LAN link interconnecting two disparate wireless LANs.

Wireless LAN Adapters

Wireless LAN adapters or wireless network interface cards (wireless NICs) are devices that individuals use with their wireless computing devices to participate in a wireless LAN. Just like APs, wireless LAN adapters are technology specific and must be carefully chosen based on the LAN requirements and the AP technology.

Per-site user and device profiling should help you estimate the number of wireless LAN adapters that you will need. The wireless LAN adapters will be different for mobile devices (for example, notebooks and PDAs) and desktop workstations. Mobile devices will generally require PC Card Interface Adapter–based wireless LAN adapters, whereas desktop workstations would typically need Peripheral Component Interconnect (PCI)–based wireless LAN adapters. Depending on the deployment, plan on using the type of wireless LAN adapters that fit your needs. Figure 7.2 shows some of the most commonly used LAN adapters.

Figure 7.2 Commonly used wireless LAN adapters.

Routers

Routers are the devices that act as a police officer guiding traffic between two or more LANs. For example, if a wireless LAN consists of 100 computers and 50 of these computers belong to department A and the other 50 to department B, assuming that the computers in the two departments do not necessarily communicate with each other as much as they communicate with computers within their department, it is a good idea to build two separate wireless LANs, A and B. A network divided this way provides higher performance as each LAN is now standalone and does not interfere with traffic from the other. If communication between networks A and B is desired, they can be interconnected using a router. The router will intelligently send data packets from LAN A to LAN B only if a data packet originating in LAN A is intended for a computer in LAN B. This routing of LAN traffic is illustrated in Figure 7.3

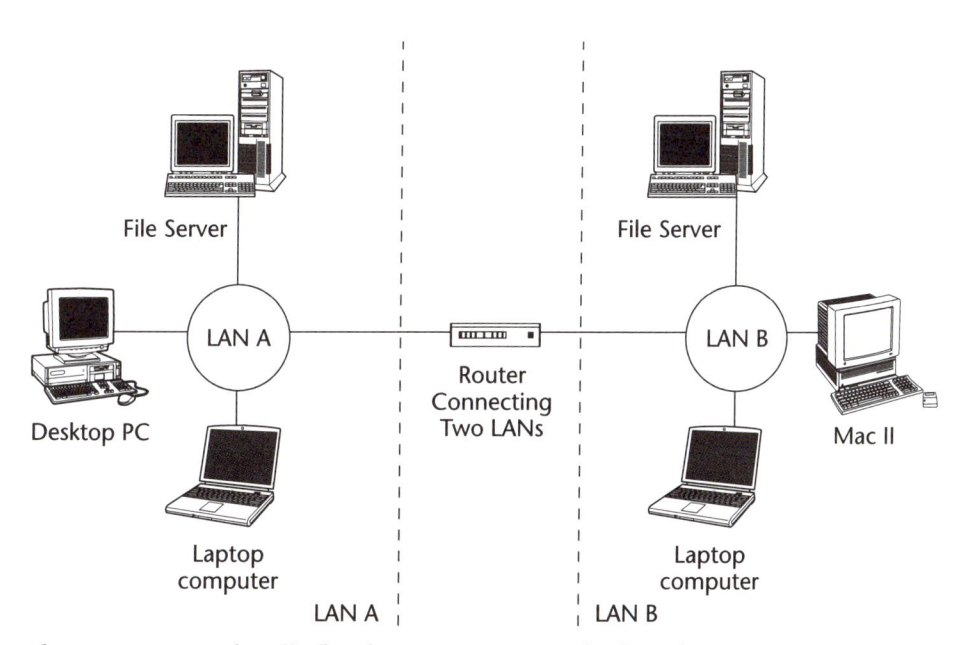

Figure 7.3 Network traffic flow between two networks through a router.

Some APs do come with built-in routers to help separate wireless LANs from wired LANs. If you are planning a standalone wireless LAN that consists of a few computers—for example, 10 or fewer—you might not need a router. However, if you are planning to interconnect a wireless LAN with a wired LAN or to the Internet, you should use a router to isolate the two LANs.

Hubs

Hubs are wired LAN devices that extend a LAN by adding more physical connection points on a LAN (that is, RJ45 LAN Jacks). You might need to plan using one or more hubs if you are planning a LAN that will contain a substantial number of wired LAN computers.

Firewalls

Firewalls are highly configurable network devices that can be programmed to allow only restricted traffic to flow through networks. The use of firewalls is extremely important if you are connecting a LAN to the Internet. You should plan on using firewall equipment if your intended LAN interconnects two or more LANs especially over the Internet.

Broadband Connectivity Devices

The broadband connectivity hardware connects a LAN to a broadband connection, for example a digital subscriber line (DSL) modem connects a LAN to the Internet. Each type of broadband connectivity requires a different type of hardware, and it can be obtained from the service provider of the broadband connection. Plan on obtaining such a device if you intend to provide broadband connectivity to your wireless LAN.

Software

You should plan on acquiring operating systems and the drivers for each type of wireless LAN adapter and the operating system that you will be using. The wireless LAN adapter manufacturers usually ship the software drivers along with the adapter, or these can be downloaded from the manufacturer Web site for most popular operating systems. When planning to use a certain type of wireless LAN adapter, make sure that the vendor of the wireless LAN does provide support and software drivers for the operating systems that you will be using. For example, if you are planning to use wireless LAN adapters with Microsoft Windows XP, make sure that the manufacturer of the wireless LAN adapter supplies the drivers and other necessary software.

Conventional Hardware Requirements for Various Deployment Scenarios

In this section, we talk about the most popular hardware requirements of home, SoHo, enterprise, and WISP deployment scenarios. This section may help you plan your particular wireless LAN hardware needs.

Home

Home deployments of wireless LANs are the simplest and often require the least hardware. You will need the following for a simple wireless home LAN:

- **Wireless LAN adapters.** You will need one wireless LAN adapter for each computer or handheld device that will be participating in a LAN.
- **Wireless LAN APs.** You will need one or more APs to provide LAN connectivity.
- **Broadband connection device, network router, and firewall.** DSL or cable modem, router, and firewall are necessary to provide a safe connection to the Internet. Many manufacturers are selling APs that have built-in routers and firewalls and can be easily plugged into the broadband connectivity hardware.
- **Wired LAN cables.** You might also need some wired LAN connectivity for connecting stationary LAN devices like network printers with the AP. Most APs normally provide direct physical LAN jacks for connecting with such devices.

Small Office, Home Office (SoHo)

The small office, home office (SoHo) deployment scenarios are similar to those of home deployments except they often contain one or more file servers and require higher security. Plan on setting up proper security and deploying file servers when establishing a wireless LAN in a SoHo environment.

Wireless Internet Service Providers (WISPs)

The wireless Internet service providers (WISPs) provide wireless Internet services at public places like coffee shops, airports, and shopping malls. WISPs normally provide these services by installing APs at public places and connecting the APs with a broadband connection like DSL. If you are planning to establish a WISP hotspot (point-of-service), you should have at least the following equipment installed at every site where you intend to provide WISP services:

Wireless LAN APs. APs to provide wireless LAN connectivity to the end-users.

Broadband connection device. WISPs only provide access to the Internet and normally do not provide any wireless LAN security. Under most circumstances, the broadband connection device is directly connected to the APs to provide a data path to the Internet.

WISP account management software. WISP services are normally not free and require a fee to access the Internet. In order to account for the service usage and authentication, WISPs normally deploy account management software that monitors usage and bills a customer credit card for the charges incurred. There are not many well-known accounting software companies that provide such software and services. Therefore, WISPs normally employ software engineers who write programs to manage the user accounts. These programs are normally installed at the WISP hotspot or at a remote site that is connected with the hotspot. A WISP deployment can be illustrated as shown in Figure 7.4.

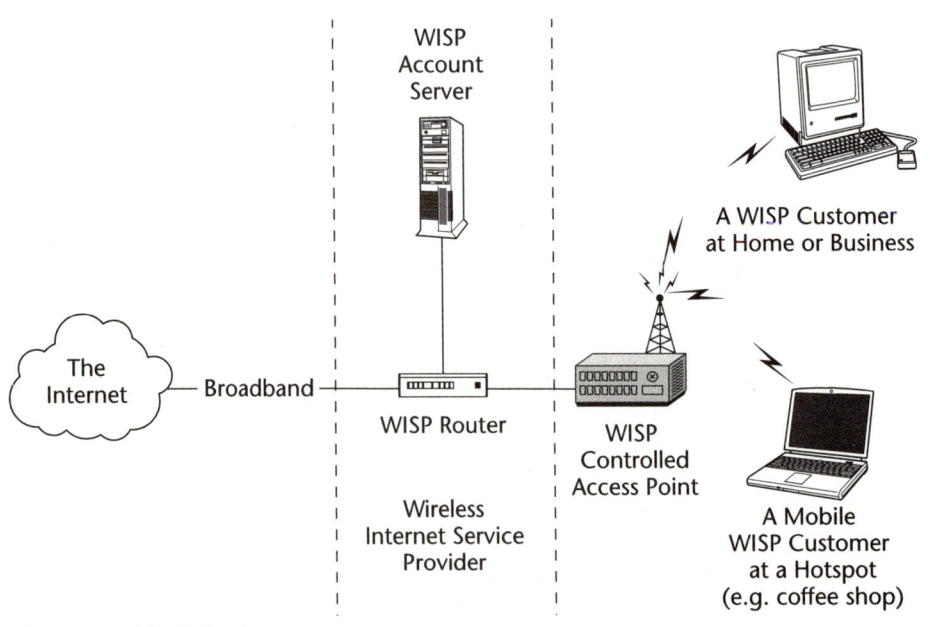

Figure 7.4 WISP deployment scenario.

Enterprise

Enterprise wireless LAN deployments are the most complicated deployment scenarios. Enterprise requirements normally include connectivity to the Internet and with other corporate branches, the support for both wired and wireless LANs, and high security. The following are important factors to consider when planning a wireless LAN for an enterprise:

- **Wireless LAN adapters.** You will need one wireless LAN adapter for each computer or handheld device that will be participating in a LAN. Plan on acquiring various different types of wireless LAN adapters as your site survey suggests.
- **Wireless LAN APs.** You will need many APs to provide wireless LAN connectivity.
- **Broadband connection device and network router and firewall.** Enterprise LANs normally have higher speed requirements than home or SoHo users and often use various different types of equipment within a network. Plan to acquire network routers and firewalls to provide adequate security that is mandated by the enterprise policies.
- **Wired LAN cables and miscellaneous hardware.** Enterprise LANs have higher data throughput requirements that cannot be achieved using the current wireless LAN technologies. Enterprise LANs almost always use wired LANs in network segments and users that require high network bandwidth, whereas the network segments and users requiring lower throughputs are connected using the wireless LAN equipment.
- **Security measures.** Wireless LANs must ensure that the best available security measures are used to protect the privacy of data traveling in a wireless LAN.

Step 6: Evaluating the Feasibility of Wireless LANs and the Return on Investment (ROI)

The planning for a wireless LAN is not complete until the network's feasibility is concretely defined. Deploying a wireless LAN might not be the best idea for all scenarios. You should carefully examine the outcome of your planning. In addition to feasibility, you might also want to estimate the return on investment (ROI) of deploying a wireless LAN. The calculation of ROI should include the costs associated with deployment and usage and the amount of money to be saved over a reasonable period of time. A positive ROI will be a good sign for wireless LAN deployment, whereas a negative ROI could become an inhibiting factor.

Step 7: Communicating the Final Plan with Higher Executives and Potential Users

The wireless LAN deployment plan should be carefully documented to present an unbiased solution that provides reasonable benefits over a wired LAN. It should address all the points discussed in this chapter along with any future upgrade options that might benefit wireless LAN deployment and protect the investment on the wireless LAN.

We suggest that you share the planning document with executives and potential users to get their opinion on your wireless LAN deployment plans.

An Example of Wireless LAN Planning: Bonanza Corporation

To understand the planning process better, let's walk through an example of wireless LAN planning at a hypothetical company called Bonanza Corporation. Following are some of our assumptions about the Bonanza Corporation:

- Bonanza has an office in San Francisco and another office in New York City.
- Bonanza has 35 employees in the New York office and 25 employees in the San Francisco office.
- Bonanza has never had a LAN before. This is the first time they are deploying any computer network.
- Ms. Leah is the IT manager at Bonanza. Ms. Shelly, the CEO of the company, asked her to build a LAN that interconnects the two offices and provides seamless and secure wireless LAN connectivity to the users at both sites.

- Leah is a computer enthusiast and an early adopter. She was using a wireless LAN at home that she built and is extremely excited to have an opportunity to build a secure wireless LAN at work.
- We further assume that Leah is a very disciplined individual and she follows the steps defined in this book to deploy a wireless LAN. She starts the process by first understanding the technologies involved, then undertaking the planning process.

Step 1: Bonanza Wireless LAN Needs

Leah estimates that the users at the two LANs, San Francisco and New York City, need at least one file server at each site; SF needs one printer, whereas NYC needs two printers; and both sites need Internet access. She further realizes that she has to provide remote workers the ability to securely connect with the corporate LANs. She decides to use virtual private network (VPN) gateways at each site to provide such connectivity over the Internet. She summarizes the wireless LAN as shown in Table 7.1.

Leah figures that she needs to plan for two separate wireless LANs, one for each office, with wired LAN extensions. She plans to attach all stationary devices and computers to the wired LAN, whereas she plans to connect mobile devices with the wireless LANs. She draws a diagram of the desired LAN as shown in Figure 7.5.

Table 7.1 LAN Needs at Bonanza Corporation

REQUIREMENT	PURPOSE
Wireless LAN connectivity	To provide staff with mobile computing devices the freedom to move about and enhance productivity.
Network File Server	To share files and documents.
Network Printers	To print documents.
Internet Access	Each site must have Internet access to interconnect LANs at both offices and to provide Web browsing services to local users.
VPN Gateways	To provide connectivity to remote users.

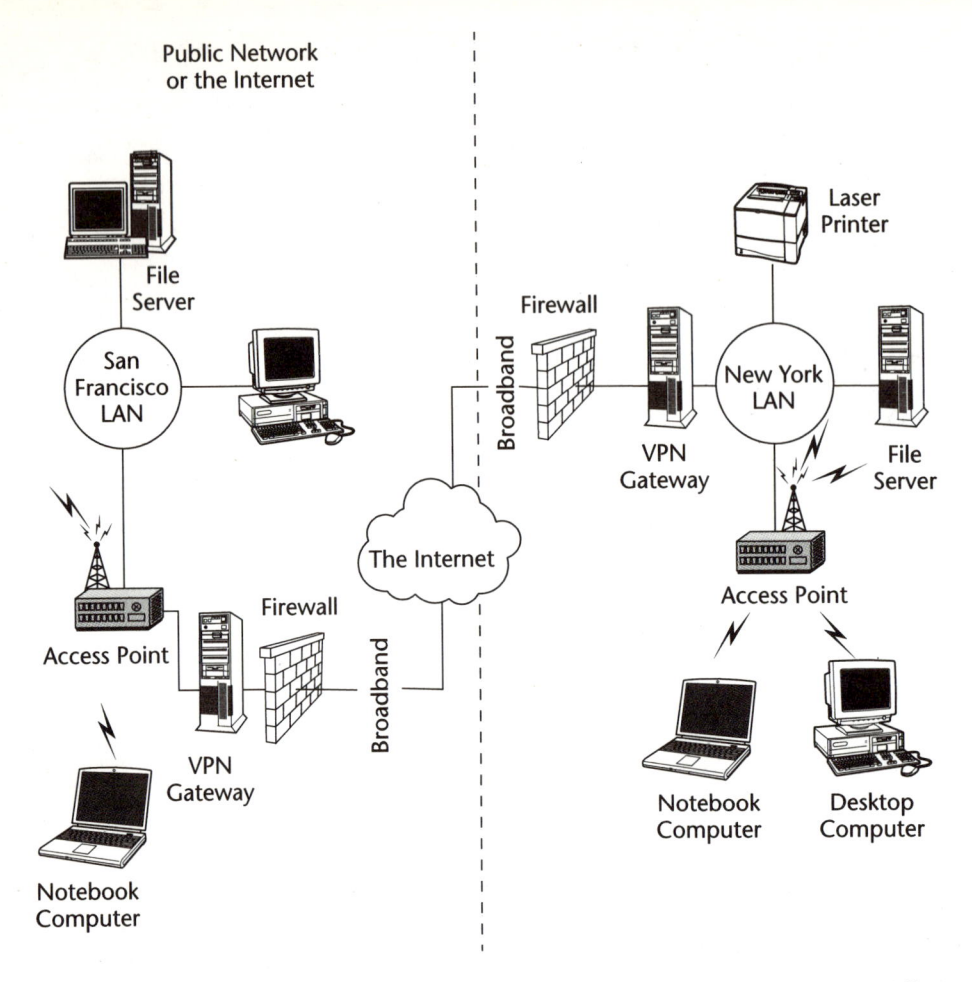

Figure 7.5 Overall wireless LAN at Bonanza Corporation.

Sample LAN at the San Francisco Office

Sample LAN at the New York City Office

Step 2: Planning the Rollout

Since there were no networks in existence at either office, Leah decides that the two networks will be rolled out in one stage, but she plans to conduct a pilot to demonstrate the system to her boss and get her approval on the deployment plans.

Step 3: Site Survey

Leah performs a site survey by visiting each office. She carries an AP with her and uses her laptop computer, equipped with a wireless LAN adapter and monitoring software, to figure out the dead spots and the best locations for the APs. She realizes that the site in New York is organized in work cubicles, whereas the San Francisco site has individual offices. She estimates that she would need fewer APs in the New York office than in the San Francisco office. She draws a site map and specifically marks the locations where she wants to install the APs and asks the facilities' coordinators to get approval from their building management. She notices that the New York office is situated in a high-rise building and needs a high level of security, whereas the office in San Francisco is situated in a Victorian house that would require less security. She is happy to use WEP encryption. She profiles the users as shown in Table 7.2 and Table 7.3.

Table 7.2 LAN Equipment Profile at San Francisco Office

DEPARTMENT	NUMBER OF USERS	COMPUTING DEVICE	NEEDS WIRELESS LAN
Sales	5	Notebooks	Yes
Engineering	10	Notebooks and PDAs	Yes
Facility Staff	2	Desktop computers	No
IT	3	Notebooks	Yes
Executives	5	Notebooks	Yes
LAN Servers and Devices (file servers and printers)	3	Servers	No
Near-term growth	5	Notebooks	Yes
Total LAN connections needed	33		

Table 7.3 LAN Equipment Profile at New York City Office

DEPARTMENT	NUMBER OF USERS	COMPUTING DEVICE	NEEDS WIRELESS LAN
Telesales	5	Desktop computers	No
Sales	15	Notebooks	Yes
Engineering	5	Notebooks and PDAs	Yes
Facility Staff	5	Desktop computers	No
IT	1	Notebook	Yes
Executives	4	Notebooks	Yes
LAN Servers and Devices (file servers and printers)	3	Servers	No
Near-term growth	5	Notebooks	Yes
Total LAN connections needed	43		

Step 4: Setting Up Requirements and Expectations

Leah had a small budget to establish the corporate LAN. She decides to establish minimum requirements and expectations for her LAN and communicates these with her boss to ensure that there are not many negative surprises when actual deployment takes place. The following were her minimum requirements and expectations:

- The proposed LAN will not be a complete wireless solution. It will be a hybrid LAN consisting of both wired and wireless LAN-based technologies.
- Only those users who have mobile needs and require relatively lower network bandwidth will be provided with wireless LAN technology. Users with fixed workstations or devices would be supplied with wired LAN connectivity solutions.
- Minimum desired speed for wireless LANs is 11 Mbps.
- There might be areas in both LANs where wireless LAN signals will be weak and the network will not perform at its best.
- Wireless LAN technologies are still evolving, and the initially deployed LANs might need to be upgraded in coming years to provide higher bandwidth and speed.

Step 5: Estimating the Required LAN Hardware and Software

After the site survey and setting up minimum requirements, Leah estimates the equipment she will need to construct the corporate LAN. She estimates the required LAN hardware and software using the knowledge she gained about the two sites during the site survey and user-profiling. Table 7.4 shows Leah's estimate for the overall corporate LAN that she is planning to deploy.

Table 7.4 Estimated LAN Hardware and Software for Bonanza Corporate LAN

EQUIPMENT	DESCRIPTION	REQUIREMENTS	QUANTITY
Wireless LAN Adapters	Network interface cards for providing wireless connectivity to the LAN devices.	At least 11 Mbps. Must be PCMCIA compliant.	58
Access Points	Wireless LAN access points facilitate LAN connectivity among the devices operating in the wireless LAN.	At least 11 Mbps with at least 300 meters' range in open space.	4
Wired LAN Adapters	For providing LAN access to stationary workstation and devices.	PCI Ethernet cards with minimum 100 Mbps of speed.	18
Router	To isolate each LAN from the Internet.	6 ports.	2
Hubs	To extend the physical wired LANs.	10 ports.	4
DSL Modem	To provide Internet access to each LAN.	DSL provider specific. Best speed within the budget.	2
Virtual Private Network Gateway and Client Software	Provides secure access to remote workers.	One with industry-standard security features and supports at least 10 simultaneous remote connections.	2
VPN Hardware for Connecting the two LANs.	The two LANs must be connected with each other over the Internet using VPN. This will consist of hardware VPN clients that talk to VPN gateways.	Industry standard. Provides one VPN connection with remote gateway.	2

Step 6: Evaluating the Feasibility of Wireless LANs and Estimating Return on Investment (ROI)

Leah knows that although she thinks that wireless LANs are the ideal solution for her organization, she still needs to convince the decision-makers with a solid understanding of the feasibility of using wireless LANs and related return on investment. Leah assesses that the following demonstrates the feasibility of using wireless LANs at Bonanza Corporation:

Enhanced mobility. Most of the LAN users are mobile professionals and commute between offices and client sites. Those who do not commute between various Bonanza facilities and client sites often need to take their mobile computing units (laptops and PDAs) to meeting rooms where they require LAN connectivity. With wired LANs, providing access to mobile professionals could be quite difficult, especially in conference rooms where wired connections might not be aesthetically appealing.

Ease of deployment and management. Wireless LANs are far easier to deploy and manage. Leah thinks that the entire deployment process will take less than one-fourth of the time that it will take to deploy a wired LAN. To upgrade the LAN, Leah's assessment is that only a change of APs and network cards will be required. No more pulling all the network cable and reinstalling it again.

Leah is comfortable with the ROI that using wireless LAN will provide. Here are some of the ROI elements she considered. (We avoid considering monetary values as they depend on labor cost and current prices of networking equipment.)

Fewer cables involved. Networking cables are often one of the costliest items in building a wired LAN. Since her plan includes limited wired LANs, Leah does not have to account for the cost of wires and labor for running the wire throughout the Bonanza offices. Leah considers this a big win for her ROI assessment.

Enhanced productivity. Leah is confident that ease of wireless LAN usage and increased mobility of staff will greatly enhance productivity. Staff would be able to share their ideas, salespeople would be able to perform live demonstrations of products, and engineers would be able to brainstorm in larger groups in offices and in conference rooms. Leah thinks that this enhanced productivity will indirectly affect the revenue generation process at Bonanza Corporation.

Step 7: Communicating the Wireless LAN Deployment Plan with Executives

Upon completion of the planning stages, Leah writes a comprehensive document detailing the outcomes of her research, the requirements she sees, and the estimated equipment that she will need to build a secure wireless LAN for Bonanza Corporation. Because such documents differ in each deployment scenario and organization, we leave this exercise up to you. However, we do encourage you to include all the information you gathered during planning.

Summary

Planning a wireless LAN is an intensive and extremely important process that requires a good understanding of networking concepts and the wireless LAN technologies. A carefully planned wireless LAN ensures proper operation upon deployment by addressing the needs of users, selecting the best fit technology, and providing a wireless LAN environment that can be extended without much change to the original deployment. Therefore, carefully planning a wireless LAN step by step is extremely important and should always be included when building a secure wireless LAN.

In the next chapter, we guide you through the steps that might help you shop for wireless LAN equipment. We talk about various networking equipment that you might need and their purposes. We also list some of the major network equipment vendors and their wireless products to give you an idea of what is currently available in the market. We also give you some shopping tips to help you choose the equipment that is right for you.

CHAPTER 8

Shopping for the Right Equipment

After planning your wireless LAN, you are now ready to shop for the equipment that you will use to build a secure wireless LAN. When building a LAN, it is extremely important to make sure to buy the best equipment within your budget that satisfies all the needs and results in a high-performance and extensible wireless LAN.

Today, shopping for wired LAN is much easier as the wired LAN technologies are very well defined. However, wireless LAN technologies and standards are still evolving, and with new standards coming out every day and new features being added, it is a good idea to be careful when investing in wireless LAN equipment. This point is so significant that we have dedicated this entire chapter to helping you understand your needs to be able to make the best decisions when purchasing wireless LAN equipment. We define shopping for wireless LAN equipment as a step-wise process: First you make your shopping list from the information that you gathered in the planning phase; then, using your knowledge of wireless LAN technologies, you compare the available products in the market with your needs to figure out the best possible match for your deployment scenario; in the third step, you seek out the lowest price for the items that you need to buy; in the fourth and final step, you actually purchase the merchandise.

Chapter 8

In this chapter, we first talk about how to shop for the components that you need to set up a wireless LAN that is based on an IEEE 802.11 standard. We also talk about some of the major vendors and their products to give you a concrete example. Finally, we talk about the places on the Internet where you can buy the wireless LAN equipment for cheap without compromising the performance of your wireless LAN.

Making Your Wireless LAN Equipment Shopping List

If you planned your wireless LAN deployment, you should have a good understanding of the items that you need to build your LAN. To make your shopping list, you should list all items as identified in the planning step on a piece of paper with your minimum requirements along with the quantities desired. For our Bonanza Corporation example, which we discussed in Chapter 7, Table 8.1 illustrates the combined needs of the two LANs that Leah will be deploying.

Table 8.1 LAN Equipment Shopping List for Bonanza Corporation

EQUIPMENT	DESCRIPTION	REQUIREMENTS	QUANTITY
Wireless LAN Adapters	Network interface cards for providing wireless connectivity to the LAN devices.	At least 11 Mbps. Must be PC Card compliant.	58
Access Points	Wireless LAN access points facilitate LAN connectivity among the devices operating in the wireless LAN.	At least 11 Mbps with at least 300 meters' range in open space.	4
Wired LAN Adapters	For providing LAN access to stationary workstation and devices.	PCI Ethernet cards with minimum 100 Mbps of speed.	18
Router	To isolate each LAN from the Internet.	6 ports.	2
Hubs	To extend the physical wired LANs.	10 ports.	4
DSL Modem	To provide Internet access to each LAN.	DSL provider specific. Best speed within the budget.	2

Explore the LAN Technologies Available in the Market

After making the shopping list, the next step is to explore the technologies available in the market that satisfy your needs. You should evaluate both the wireless LAN technologies that you need and the wired LAN technologies that you will be deploying in your LAN.

Wireless LAN Technologies

This book focuses on wireless LAN based on the IEEE 802.11 technologies. In this section, we only talk about the currently available wireless LAN technologies that use the 802.11 standard. Today, two major wireless LAN technology-based equipments are available, each based on an IEEE standard. These standards are the IEEE 802.11b and IEEE 802.11a.

The IEEE 802.11b standard operates at speeds up to 11 Mbps. Following are the highlights of 802.11b. 802.11a standard devices operate at up to 54 Mbps (see Chapter 3 for more information on 802.11 standards). It is important to remember that the two IEEE standards, 802.11b and 802.11a, are incompatible with each other. A good idea is to always build a comparison matrix to visualize the differences in the equipment properties that concern you the most. Table 8.2 shows a sample matrix that compares the basic properties of the 802.11b standard with the 802.11a.

The devices based on 802.11b arrived on the market earlier than 802.11a and are less expensive than 802.11a-based devices. However, 802.11a devices provide higher speeds, which might be critical in certain deployment scenarios.

Wired LAN Ethernet Equipment Technologies

Since this book is focused on wireless LANs that are built on the 802.11 standard, also known as wireless Ethernet, we limit our discussion on wired LANs to Ethernet-based technologies. (See Table 8.3.)

Table 8.2 802.11b Compared with 802.11a

TECHNOLOGY STANDARD	MAXIMUM SPEED	FREQUENCY BAND	GEOGRAPHIC RANGE
802.11b	11 Mbps	2.4 GHz	300 meters
802.11a	54 Mbps	5.4 GHz	200 meters

Table 8.3 Common Ethernet Standards

TECHNOLOGY STANDARD	MAXIMUM SPEED
Ethernet	10 Mbps
Fast Ethernet	100 Mbps
Gigabit Ethernet	1 to 2 Gbps

As is the case for wireless LAN technologies, the higher the speed of a network, the more expensive the equipment to build a network using that technology. Today, Fast Ethernet LANs are very common, and Gigabit Ethernet LANs are slowly being adopted. Under most circumstances, Gigabit Ethernet LANs are operable with Fast Ethernet. We suggest that you use Gigabit Ethernet adapters with devices that require high speed, and Fast Ethernet in computers that do not have high-speed LAN requirements. For example, file servers should be installed with Gigabit Ethernet adapters, and desktop computers should be supplied with the Fast Ethernet adapters.

Virtual Private Network (VPN) Gateways and Clients

Virtual private networks are becoming extremely popular. Most enterprise LANs deploy VPN gateways to allow remote workers secure access to the enterprise LAN. VPNs consist of two basic components: a VPN gateway, which is normally a hardware device and resides at the physical site, and the VPN client, which is normally a software application program and is installed on the user computers. When exploring VPNs, you must ensure that the security algorithms supported by the VPN provide adequate security for your needs.

Remote Authentication Dial-in User Service (RADIUS) Server

The Remote Authentication Dial-in User Service (RADIUS) server is used to authenticate remote clients. The 802.1X authentication protocol provides support for RADIUS servers. 802.1X will be available in 802.11-based devices that will be released in the near future. RADIUS servers are available from many different vendors. There are also many open-source RADIUS implementations that provide comparable services and can be obtained without any cost. Some well-known RADIUS implementations are listed in Table 8.4.

Table 8.4 Popular RADIUS Server Vendors

VENDOR	PRODUCT
Nortel Networks, Inc.	Nortel Networks RADIUS Server
A Free Implementation of RADIUS server distributed under the GNU GPL license	Cistron RADIUS Server

If you are interested in using the 802.11-compliant devices that provide security through the use of the 802.1X standard, you should plan on purchasing or acquiring a RADIUS server.

Wireless LAN Supporting Operating Systems

Wireless LAN adapters require software drivers for the operating system (OS) that they need to be operating under. For example, if you have a laptop with Windows XP and a wireless LAN adapter from Cisco Systems, you will need software drivers for Windows XP from Cisco Systems. When choosing a wireless LAN adapter, you must always ensure that the vendor supports the OS you intend to use the adapter with. Most 802.11-compliant device manufacturers support the following operating systems:

- Microsoft Windows XP, Windows 2000, Windows 98, and Windows ME
- Microsoft Windows NT (version 4.0 and above), Windows 2000 Advanced Servers, and Windows XP Servers
- Linux operating system
- Palm operating system
- Microsoft PocketPC operating system

If your LAN has computing devices that utilize operating systems not listed here, you should certify from the wireless LAN adapter manufacturer that your operating system is supported by the device you are interested in buying. Otherwise you should plan on buying an operating system that is supported by the vendor or choose a different wireless LAN adapter.

Major 802.11 Equipment Vendors and Their Products

Today, over 25 big vendors are providing 802.11-based wireless LAN equipment. These companies range from some of the biggest names in the networking industry to small hardware manufacturers. In this section, we list some of the well-known companies and their products to give you a baseline understanding of the products available today. Following are some of the basic parameters that we list for each product to help you choose the right vendor and product:

- **Data rates**. Data rates are the speeds at which certain LAN equipment operate. Different models have different speeds. For example, 802.11b has a maximum speed of 11 Mbps.

- **Operating range**. The operating range is normally expressed as maximum number of feet LAN equipment can operate with or without degradation of performance. Different models of the same product, standard, or vendor may offer different ranges at different prices. You should carefully select a LAN device to ensure that it will fit your needs.

- **Models**. Different models normally come with different features. Remember to write down the model number of each product that you like, as the external packaging of two very different devices might appear identical.

- **Encryption key length**. Encryption keys provide security to wireless LANs. Devices that use longer encryption keys are supposed to provide higher security.

- **Security protocols**. Security protocols provide the security mechanism that is used to secure a LAN. WEP and 802.1X are examples of security protocols that are used in wireless LANs.

- **Remote configuration**. Remote configuration normally refers to a feature that many LAN devices provide that enables a LAN manager to configure and manage a LAN device from a remote location or from his or her desk. This feature allows the expensive equipment to reside in a physically secured location, and the administrator does not have to enter into the secure location to configure or manage the hardware. This feature also enables administrators and network managers to manage a LAN device from a geographically separated site.

Cisco Systems

Cisco was founded in 1984 by a group of computer scientists from Stanford University. Since the company's inception, Cisco engineers have been prominent in advancing the development of wired and wireless network technologies. The company's tradition of innovation continues today with Cisco creating leading products and key technologies that will make the Internet more useful and dynamic in the years ahead. These technologies include advanced routing and switching, voice and video over IP, optical networking, wireless, storage networking, security, broadband, and content networking. More information on Cisco Systems can be obtained from their Web site at http://www.cisco.com. Following are some of the wireless LAN products that Cisco Systems currently ships.

All Cisco Aironet 350 Series client adapters and access points are IEEE 802.11b compliant. The Cisco Aironet 350 Series was the first product to deliver a wireless LAN solution that offered centralized 802.1X-based security.

802.11b Products

Table 8.5 shows the major Cisco products based on 802.11b.

Table 8.5 Major 802.11b Products

CISCO AIRONET 350 SERIES ACCESS POINTS	
FEATURES	**DESCRIPTION**
Data Rates	1, 2, 5.5, and 11 Mbps
Operating Range	Indoor: 130 ft (39.6 m) @ 11 Mbps 350 ft (107 m) @ 1 Mbps Outdoor: 800 ft (244 m) @ 11 Mbps 2000 ft (610 m) @ 1 Mbps
Models	AIR-AP352E2C, the standard AP
	AIR-AP352E2R-A-K9, the rugged AP configured for operation in most of the Americas
	AIR-AP352E2R-E-K9, the rugged AP configured for operation in most of Europe and Singapore
	AIR-AP352E2R-J-K9, the rugged AP configured for operation in Japan
Encryption Key Length	128 bit
Security Protocols	IEEE 802.1X (proposal includes EAP and RADIUS) and IEEE 802.11 WEP (Wired Equivalent Privacy)
Remote Configuration	Telnet, HTTP, FTP, TFTP, and SNMP

(continues)

Table 8.5 Major 802.11b Products *(Continued)*

CISCO AIRONET 350 WIRELESS LAN ADAPTER	
FEATURES	**DESCRIPTION**
Data Rates	1, 2, 5.5, and 11 Mbps
Operating Range	Indoor: 130 ft (39.6 m) @ 11 Mbps 350 ft (107 m) @ 1 Mbps Outdoor: 800 ft (244 m) @ 11 Mbps 2000 ft (610 m) @ 1 Mbps
Models	AIR-PCM35x: PC Card (PCMCIA) Type II AIR-PCI351X: peripheral component interconnect (PCI) Bus.
Encryption Key Length	128 bit
Security Protocols	Security IEEE 802.1X (proposal includes EAP and RADIUS) and IEEE 802.11 WEP (Wired Equivalent Privacy)
Authentication	Extensible Authentication Protocol (EAP)

Agere Systems/ORiNOCO

ORiNOCO is one of the first manufacturers of wireless LAN devices based on 802.11. ORiNOCO is also known to provide support for more operating systems than any other hardware vendor. More information on ORiNOCO can be obtained from http://www.orinocowireless.com.

802.11b Products

Table 8.6 shows the major ORiNOCO products based on 802.11.

Table 8.6 The major ORiNOCO products based on 802.11

ORINOCO AP-200 ACCESS POINT	
FEATURES	**DESCRIPTION**
Data rates	1, 2, 5.5, and 11 Mbps
Operating Range	Indoor: 80 ft @ 11 Mbps 165 ft @ 1 Mbps Outdoor: 525 ft @ 11 Mbps 1750 ft @ 1 Mbps
Models	ORiNOCO AP-200 Access Point
Encryption Key Length	64 bit and 128 bit
Security Protocols	IEEE 802.11 WEP (Wired Equivalent Privacy)
Remote Configuration	HTTP (via Web browser), SNMP, Telnet, and TFTP

ORINOCO WORLD PC CARD	
FEATURES	**DESCRIPTION**
Data Rates	1, 2, 5.5, and 11 Mbps
Operating Range	Indoor: 80 ft @ 11 Mbps 165 ft @ 1 Mbps Outdoor: 520 ft @ 11 Mbps 1750 ft @ 1 Mbps
Models	ORiNOCO World PC Card
Encryption Key Length	64 bit and 128 bit
Security Protocols	IEEE 802.11 WEP (Wired Equivalent Privacy)

5-GHz Migration Products Based on 802.11b

The migration products normally include support for both existing and upcoming protocol standards. ORiNOCO AP-2000 Access Points is a migration product and provides support for both 802.11a and 802.11b standards through CardBus interface. Table 8.7 shows the major 5-GHz migration products based on 802.11b.

Linksys

Linksys was founded in 1988 and has become the fastest growing networking vendor in the distribution channel, which caters to small/medium businesses, corporate workgroups, and enterprise environments through value added resellers (VARs) and catalogs.

Table 8.7 5-GHz Migration Products Based on 802.11b

ORINOCO AP-2000 ACCESS POINT	
FEATURES	**DESCRIPTION**
Data Rates	1, 2, 5.5, and 11 Mbps - Allows up to two CardBus card installation for enhanced performance
Operating Range	Indoor: 80 ft @ 11 Mbps / 165 ft @ 1 Mbps / Outdoor: 525 ft @ 11 Mbps / 1750 ft @ 1 Mbps
Models	ORiNOCO AP-2000 Access Point
Encryption Key Length	64 bit and 128 bit
Security Protocols	IEEE 802.1X (includes EAP-TLS and RADIUS) and IEEE 802.11 WEP (Wired Equivalent Privacy)
Remote Configuration	Telnet, HTTP, FTP, TFTP, and SNMP

802.11b Products

Table 8.8 shows some of the major Linksys wireless products based on 802.11b.

Table 8.8 Linksys Wireless Products Based on 802.11b

NETGEAR MA401 802.11B WIRELESS PC CARD

FEATURES	DESCRIPTION
Data Rates	1, 2, 5.5, and 11 Mbps
Operating Range	Indoor: 175 ft @ 11 Mbps 500 ft @ 1 Mbps Outdoor: 835 ft @ 11 Mbps 1650 ft @ 1 Mbps
Models	NETGEAR MA401 802.11b Wireless PC Card
Encryption Key Length	40 bit and 128 bit
Security Protocols	IEEE 802.11 WEP (Wired Equivalent Privacy)

INSTANT WIRELESS NETWORK ACCESS POINT

FEATURES	DESCRIPTION
Data Rates	1, 2, 5.5, and 11 Mbps
Operating Range	Indoor: 164 ft @ 11 Mbps 492 ft @ 1 Mbps Outdoor: 820 ft @ 11 Mbps 1640 ft @ 1 Mbps
Models	Instant Wireless Network Access Point
Encryption Key Length	64 bit and 128 bit
Security Protocols	IEEE 802.11 WEP (Wired Equivalent Privacy)
Remote Configuration	HTTP (via Web browser) and simple network management protocol (SNMP)

NetGear

NetGear was founded in January 1996 by Patrick Lo and Mark Merrill, and operated as a wholly owned subsidiary of Bay Networks, which was purchased by Nortel Networks in August 1998. The company continued to operate as a wholly owned subsidiary of Nortel until it was spun off as a separate company in March 2000. Today, NetGear produces network connectivity products that include wireless, Ethernet network interfaces, hubs, and switches. Following are some of the NetGear IEEE 802.11 products.

802.11b Products

Table 8.9 shows the major NETGEAR products based on 802.11b.

Table 8.9 NetGear Products Based on 802.11b

NETGEAR MA401 802.11B WIRELESS PC CARD	
FEATURES	**DESCRIPTION**
Data Rates	1, 2, 5.5, and 11 Mbps
Operating Range	Indoor: 175 ft @ 11 Mbps 500 ft @ 1 Mbps Outdoor: 835 ft @ 11 Mbps 1650 ft @ 1 Mbps
Models	NetGear MA401 802.11b Wireless PC Card
Encryption Key Length	40 bit and 128 bit
Security Protocols	IEEE 802.11 WEP (Wired Equivalent Privacy)

NETGEAR MR314 802.11B CABLE/DSL WIRELESS ROUTER WITH ACCESS POINT	
FEATURES	**DESCRIPTION**
Data Rates	1, 2, 5.5, and 11 Mbps
Operating Range	Indoor: 175 ft @ 11 Mbps 500 ft @ 1 Mbps Outdoor: 835 ft @ 11 Mbps 1650 ft @ 1 Mbps
Models	NetGear MR314 802.11b Wireless Access Point

Table 8.9 (Continued)

NETGEAR MR314 802.11B CABLE/DSL WIRELESS ROUTER WITH ACCESS POINT	
FEATURES	DESCRIPTION
Encryption Key Length	64 bit, 128 bit, and 152 bit
Security Protocols	IEEE 802.11 WEP (Wired Equivalent Privacy)
Remote Configuration	HTTP (via Web browser)

802.11a Products

Table 8.10 shows the major NetGear products based on 802.11a.

Table 8.10 NETGEAR Products Based on 802.11a

NETGEAR HA501 802.11A WIRELESS PC CARD	
FEATURES	DESCRIPTION
Data Rates	6, 9, 12, 24, 36, 48, 54, and 72 Mbps
Operating Range	Indoor: 60 ft @ 54 Mbps 300 ft @ 6 Mbps Outdoor: 100 ft @ 54 Mbps 1200 ft @ 4 Mbps
Models	NetGear HA501 802.11a Wireless PC Card
Encryption Key Length	64 bit, 128 bit, and 152 bit
Security Protocols	IEEE 802.11 WEP (Wired Equivalent Privacy)

NETGEAR HE102 802.11A WIRELESS ACCESS POINT	
FEATURES	DESCRIPTION
Data Rates	6, 9, 12, 24, 36, 48, 54, 72, and 100 Mbps
Operating Range	Indoor: 60 ft @ 54 Mbps 300 ft @ 6 Mbps Outdoor: 100 ft @ 54 Mbps 1200 ft @ 4 Mbps
Models	NetGear HE102 802.11a Wireless Access Point

(continues)

Table 8.10 NETGEAR Products Based on 802.11a *(Continued)*

NETGEAR HE102 802.11A WIRELESS ACCESS POINT	
FEATURES	**DESCRIPTION**
Encryption Key Length	64 bit, 128 bit, and 152 bit
Security Protocols	IEEE 802.11 WEP (Wired Equivalent Privacy)
Remote Configuration	HTTP (via Web browser)

Xircom/Intel Corporation

Intel Corporation acquired Xircom in March 2001. Xircom, now Intel, was always known for quality remote-connectivity products. Currently, Intel is shipping the Xircom products as well as new Intel branded wireless LAN products based on the IEEE 802.11 standards.

802.11b Products

Table 8.11 shows the major Xircom products based on 802.11b.

Table 8.11 Xircom Products Based on 802.11b

XIRCOM WIRELESS ETHERNET ACCESS POINT	
FEATURES	**DESCRIPTION**
Data Rates	1, 2, 5.5, and 11 Mbps
Operating Range	Indoor: 130 ft @ 11 Mbps 350 ft @ 1 Mbps Outdoor: 800 ft @ 11 Mbps 2000 ft @ 1 Mbps
Models	APWE1120
	Xircom Wireless Ethernet Access Point
Encryption Key Length	40 bit
Security Protocols	IEEE 802.1X (proposal includes EAP and RADIUS) and IEEE 802.11 WEP (Wired Equivalent Privacy)

Shopping for the Right Equipment 181

Table 8.11 *(Continued)*

XIRCOM WIRELESS ETHERNET ACCESS POINT	
FEATURES	**DESCRIPTION**
Remote Configuration	Telnet, HTTP, FTP, TFTP, and SNMP

INTEL PRO/WIRELESS 2011B LAN ACCESS POINT	
FEATURES	**DESCRIPTION**
Data Rates	1, 2, 5.5, and 11 Mbps
Operating Range	Indoor: 130 ft @ 11 Mbps 350 ft @ 1 Mbps Outdoor: 800 ft @ 11 Mbps 2000 ft (610 m) @ 1 Mbps
Models	APWE1120
	Xircom Wireless Ethernet Access Point
Encryption Key Length	40 bit
Security Protocols	IEEE 802.1X (proposal includes EAP and RADIUS) and IEEE 802.11 WEP (Wired Equivalent Privacy)
Remote Configuration	HTTP (via Web browser), Telnet, SNMP, FTP, and Intel feature to do bulk configuration to many access points

XIRCOM CREDITCARD WIRELESS ETHERNET ADAPTER	
FEATURES	**DESCRIPTION**
Data Rates	1, 2, 5.5, and 11 Mbps
Operating Range	Indoor: 100 ft @ 11 Mbps 300 ft @ 1 Mbps Outdoor: 400 ft @ 11 Mbps 1500 ft @ 1 Mbps
Models Protocols	CWE1130
	CreditCard Wireless Ethernet adapter
Encryption Key Length	128 bit
Security Protocols	IEEE 802.11 WEP (Wired Equivalent Privacy)

(continues)

Table 8.11 Xircom Products Based on 802.11b *(Continued)*

INTEL PRO/WIRELESS 2011B LAN ADAPTERS	
FEATURES	DESCRIPTION
Data Rates	1, 2, 5.5, and 11 Mbps
Operating Range	Indoor: 100 ft @ 11 Mbps 300 ft @ 1 Mbps Outdoor: 400 ft @ 11 Mbps 1500 ft @ 1 Mbps
Models	Intel PRO/Wireless 2011B LAN Adapters Intel PRO/Wireless 2011b LAN PCI Adapter
Encryption Key Length	64 bit, 128 bit
Security Protocols	IEEE 802.11 WEP (Wired Equivalent Privacy)

802.11a Products

Table 8.12 shows the major Xircom products based on 802.11a.

Table 8.12 Xircom Products Based on 802.11a

INTEL PRO/WIRELESS 5000 LAN ACCESS POINT	
FEATURES	DESCRIPTION
Data Rates	1, 2, 5.5, and 11 Mbps
Operating Range	Indoor: 40 ft @ 54 Mbps 300 ft @ 6 Mbps Outdoor: 100 ft @ 54 Mbps 1000 ft @ 4 Mbps
Models	Intel PRO/Wireless 5000 LAN Access Point
Encryption Key Length	64 bit, 128 bit
Security Protocols	IEEE 802.1X (proposal includes EAP and RADIUS) and IEEE 802.11 WEP (Wired Equivalent Privacy)
Remote Configuration	HTTP (via Web browser), Telnet, SNMP, FTP, and Intel feature to do bulk configuration to many access points

Table 8.12 *(Continued)*

INTEL PRO/WIRELESS 5000 CARDBUS ADAPTER	
FEATURES	**DESCRIPTION**
Data Rates	54, 48, 36, 24, 18, 12, 9, 6 Mbps
Operating Range	Indoor: 40 ft @ 54 Mbps 300 ft @ 6 Mbps Outdoor: 100 ft @ 54 Mbps 1000 ft @ 4 Mbps
Models	Intel PRO/Wireless 5000 LAN Product Family
Encryption Key Length	64 bit, 128 bit
Security Protocols	IEEE 802.11 WEP (Wired Equivalent Privacy)

Decide Your Shopping Parameters

Shopping for wireless LAN equipment could become a nightmare if you do not know your shopping parameters. Unless you are supplied with unlimited amounts of money, you should always set up a budget and a set of requirements that will constrain your shopping. You should pay special attention to the following:

- Ensure that the features important to your deployment are present in the wireless LAN equipment you are about to purchase. These features must include the speed, the security features, and interoperability with other LAN equipment that you might already have. For example, if your desired speed is 54 Mbps, you should look into buying 802.11a-compliant equipment. If you require high security in your wireless equipment, you should seek wireless LAN equipment that complies with the 802.1X security standards. Likewise, if you already have a wireless LAN infrastructure, you should purchase equipment that will interoperate with your existing infrastructure.

- Do you need equipment from a well-known vendor? Many organizations have IT policies where all IT equipment must be purchased from well-known vendors. If you choose not to follow this policy, you might be able to get the same quality from a relatively unknown vendor.

- Do you need warranty? One reason to buy equipment from a well-known vendor is the warranty that comes with the products. If warranty is a preferred criterion for you, you should stick with your policy, otherwise you could compromise your warranty protection with cheaper equipment that provides almost the same service.

Shopping for LAN Equipment

The competition ignited by the Internet in the marketplace is no secret. You can often find out about the features and pricing of a new product before it is released. While this helps you get lower prices, it may also get very confusing. When shopping, you should choose the means you are most comfortable with. In this section, we talk about the most prevalent options of buying LAN equipment.

Shopping on the Internet

When shopping for LAN equipment over the Internet, you can often buy items directly from the vendor or manufacturer's Web site, from online stores that sell computers and LAN equipment, from an online-auction Web site, or by using online shopping tools that include price comparisons on the Internet that list the prices from various vendors.

Online Stores

The following are some of the best-known online stores, comparison tools, and auction sites:

PC Connection. PC Connection, Inc. is a leading direct marketer of business computing solutions. We had the best experience with PC Connection and are inclined to list it as a top Web site. PC Connection's Web site is www.pcconnection.com.

Gateway. Gateway Computers provides great service and amazing prices. Gateway can be reached at www.gateway.com.

PC Mall. Another famous online retailer. PC Mall has a good selection of wireless equipment. PC Mall can be reached at www.pcmall.com.

Comparison Tools

We recommend that all readers try the shopping comparison tools before buying any equipment from any vendor. The comparison tools provide you with prices from vendors, their current inventory, and reviews on products.

CNET - Shopper. This provides the best online comparison services on the Internet. Besides price comparison, you can also find reviews by users that can help you decide whether equipment suits your needs. Their Web site is www.shopper.com.

StoreRunner. www.storerunner.com is also a comparison Web site that can help you compare prices, but also helps you locate a local store that has the prices you are interested in.

Online Auction Sites

Auction sites often provide you with the ability to obtain equipment at much cheaper prices than you can find at regular stores. However, purchases through auction may have the following disadvantages:

Warranty. Purchases from auction might not be covered by warranty.

Return. Often you cannot return merchandise when you purchase from an auction.

Available quantity. Items on sale at auctions are normally available in very limited quantities. If you need a substantial quantity of the same kind of equipment, you might want to avoid purchasing items through auctions.

Frauds. Online auctions may also result in frauds in which you might not receive what you expected.

We suggest that you use auctions to purchase LAN equipment if your requirements are small and you can mix and match a variety of products in your LAN. The following are some of the most popular auction sites:

EBay. This Web site is considered one of the best auction-related sites on the Internet. EBay provides secure transactions through a service called Billpoint. EBay can be reached at: www.ebay.com.

Ubid. Ubid is a reseller of a variety of electronic items including wireless LAN products. Since Ubid is a direct reseller, you might be more confident when bidding for items in an auction. Ubid can be reached at www.ubid.com.

Shopping Using Mail-Order Catalogs

Mail-order catalogs were really big for computer hardware sales during the late 1980s through the mid-1990s. Today, mail-order catalog companies conduct most of their business through their Web sites; still they continue to maintain their mail catalogs. Mail catalogs often list items that are moving relatively slowly and can be bought for cheaper prices. The following are some of the commonly known mail-order catalogs:

MicrowareHouse. This is one of the oldest mail-order catalog vendors in the United States. Microwarehouse can be reached at www.microwarehouse.com.

Gates/Arrow. Formerly known as Gates/FA is one of the biggest mail-order computer hardware and supplies company. Gates caters to high-end, high-volume customers. More information on Gates/FA can be obtained from their Web site at: www.gatesarrow.com.

Shopping at a Local Computer Hardware or Office Supply Store

Most office supply stores today sell both the wired and the wireless LAN equipment. However, the selection might be limited based on the store. Still, buying equipment directly from a store saves time and shipping charges that you might end up paying when buying equipment over the Internet. Following are some of the major computer hardware and office supply stores that carry wireless equipment.

Office Depot. Office Depot, Inc., founded in 1986, is the world's largest seller of office products and an industry leader in every distribution channel, including stores, direct mail, contract delivery, the Internet, and business-to-business electronic commerce. Office Depot has 824 Office Depot superstores in 44 states and the District of Columbia. Besides physical locations, Office Depot also has a Web site at www.officedepot.com that can provide information on store locations and online prices.

Fry's Electronics. Fry's Electronics is one of the fastest growing electronics superstores in the United States. It has over 16 stores. Prices at Fry's stores are often comparable to what you will find on the Internet. To locate a Fry's Electronics store in your neighborhood, try Fry's Web site at www.frys.com.

Staples and OfficeMax. These two are also major providers of office supplies and can be reached at www.staples.com and www.officemax.com, respectively.

Shopping Tips

Following are some suggested shopping tips. Above all, be sure to purchase equipment from the vendor with whom you are most comfortable.

Try buying equipment from a minimum number of vendors. This is important because sometimes vendors add extra functionality to their product, which provides extra features, but results in a device that is incompatible with the industry standard and the other devices that you might have. Interoperability problems are often hard to detect and can result in major network failures during deployment and operation. Since most vendors test interoperability among their own products, buying wireless LAN equipment from one vendor typically works the best.

Do not buy equipment that is based on obsolete or near-obsolete technologies. While shopping for the equipment, you may come across equipment that seems cheaper and from a better-known vendor than you could afford. In a situation like this, you should carefully examine the equipment and make sure that you are not buying something that will become obsolete in the near future, would be slower in speed, or is not supported by computer operating systems you plan to use.

Use the Internet comparison shopping tools before making shopping decisions. Comparison shopping tools are a blessing to Internet shoppers. Most sites show you the current prices at various vendors, user reviews, and you can often ask questions that you might have about a certain product.

Be careful when you purchase items online. Do not give out your credit card information to a Web site that you may suspect is a rogue site. When making a purchase online, you should make sure that the company you are about to buy the equipment from physically exists within the jurisdiction of your country or state. You should look for their credibility over the Internet to ensure that you are buying products from a real seller.

Summary

Shopping for the right equipment when deploying a wireless LAN could become quite confusing. The best approach to shop for wireless LAN equipment is to first plan the LAN, and then carefully make your shopping list and explore the market for available options. Finally, purchase the LAN equipment from the vendor that offers you the best deal for your money.

In the next chapter, we walk through the steps in provisioning a wireless LAN. We hope that the next chapter helps you understand the entire process of deploying a wireless LAN. We first identify various LAN components that you might acquire to build a wireless LAN, then we set up and test wireless LANs in both infrastructure and ad-hoc modes.

Equipment Provisioning and LAN Setup

Effortless, out-of-the-box wireless networking is one of the real gems of 802.11-based wireless LANs. The credit for simplified LAN installation and configuration goes to the cable-free nature of wireless LANs and standardization of LAN technologies. The equipment provisioning for wireless LANs involves setting up wireless LAN services and user-computing devices so that a wireless LAN can be used. Depending on the deployment type and the mode of operation, the steps involved in provisioning a wireless LAN vary. In this chapter, we talk about steps necessary to set up a wireless LAN for the two basic operation modes of 802.11-based wireless LANs: the 802.11 ad-hoc and the 802.11 infrastructure modes. We explain the process of installing wireless LAN adapters and access points. We also discuss how to connect a wireless LAN to the Internet using the ORiNOCO RG-1000 Residential Gateway AP, and the possible AP configurations to provide optimum wireless LAN performance.

Before We Start

Wireless LAN equipment is fragile and care must be taken when installing it. We strongly recommend that you carefully read all manufacturer's documentation before proceeding to install wireless equipment. Incorrect or improper installation may result in total destruction of the LAN device and/or the computing device, can cause fire, or may result in an electric shock leading to possible serious health problems. If you are not comfortable installing wireless LAN equipment yourself, you should contact a certified technician to have it done for you. If you choose to install the equipment yourself, we ask that you make sure to use extreme caution when handling and installing it. The following are good general guidelines for installing computer components:

1. Always disconnect electric power to the computers before you open them or add an adapter to them. Making sure that a computer is not supplied with electric power when you are installing a device reduces the risk of electric shock. In addition, most devices may burn out if they are installed in a computer while it is powered up.

2. Never touch the open electric components with your bare hands. Electronic components are very sensitive to static charge. If your body has static charge, it can damage the wireless LAN adapter.

3. Never attempt installation of any electrical device if you are not comfortable doing it.

4. If a device requires a power adapter, use only the power adapter that comes with the device or is certified to work with the device. Using an incorrect or improper power adapter can damage the device.

In the sections to follow, we define how to perform physical installation of both wireless LAN APs and adapters for infrastructure and ad-hoc modes.

Identifying the Wireless LAN Components

The first step toward setting up a wireless LAN involves proper identification of the wireless LAN equipment. A basic wireless LAN operating in infrastructure mode consists of wireless LAN adapters and wireless LAN access points (APs), whereas a wireless LAN operating in ad-hoc mode consists of only wireless LAN adapters. Let's walk through the steps to identify these different components.

Wireless LAN Adapters

Wireless LAN adapters are electronic devices that make up the client portion of a wireless LAN. All devices in a wireless LAN must be installed with a wireless LAN adapter. Following are the three basic forms of wireless LAN adapters that are available today.

PC Card Adapter Wireless LAN Adapters

The PC Card (formerly known as PCMCIA) standard originated in the early 1990s. It defines a small form factor (equivalent to the size of a credit card but thicker) computer hardware device. Most mobile devices including laptops, notebooks, and PDAs are equipped with PCMCIA-standard slots that accept the hardware devices that are compliant with the PC Card standard.

Today, 802.11b- and 802.11a-based PC Card devices are easily available. These devices come with their software drivers and have a built-in antenna. Figure 9.1 shows an example of a PC Card–based wireless LAN adapter.

CardBus standard, in layman terms, is a 32-bit version of PCMCIA; it is also backwards compatible with PCMCIA. Therefore, if your computer has a CardBus slot, it will work with a PC Card. But if you have a PC Card slot, it will not work with a CardBus card.

Figure 9.1 PC Card wireless LAN adapter.

Compact Flash (CF) Wireless LAN Adapters

Compact flash (CF)–based wireless LAN adapters are relatively new to the market. CF is a hardware interface standard that is smaller than PC Card and is widely used in handheld computers and PDAs. The CF connectors are similar to PC Card, and CF devices can also be used in PC Card slots using an extension device called CF jacket. Mobile users, who have both a laptop computer and a PDA, often buy devices that are CF compliant for use with their PDAs; they can also use the same CF device with their laptops using CF jacket so that they do not have to buy two separate wireless LAN adapters. An example of a wireless LAN adapter is shown in Figure 9.2.

Peripheral Component Interconnect (PCI) Wireless LAN Adapters

New hardware functionality is added to desktop computers using hardware devices that comply with the Peripheral Component Interconnect (PCI) standard. These devices contain electronic circuits on the printed circuitboards and have connectors on one edge. An example of a PCI wireless LAN adapter is shown in Figure 9.3.

Figure 9.2 A wireless LAN adapter based on the Compact Flash technology.

Figure 9.3 PCI wireless LAN adapter.

Wireless LAN Access Points (APs)

Wireless LAN access points (APs) are wireless LAN devices that make up the central portion of a wireless LAN that operate in the infrastructure mode. All LAN traffic in a wireless LAN operating in infrastructure mode must go through an access point (see Chapters 2 and 3 for more information on wireless LAN access points). Most wireless LAN APs are standalone devices and do not need to be installed or physically interfaced with a computer. Physically, APs normally have one or two antennas, small light indicators showing the status of the AP, a power jack for providing the electric current necessary to operate the device, and a few wired LAN jacks, typically RJ45, to connect the AP to a wired LAN. Figure 9.4 shows a wireless LAN AP.

Wireless LAN Antennas

Most wireless LAN devices like APs and PC Card adapters come with built-in antennas. However, if you feel that you are not receiving a strong enough signal using regular antennas, often you can add external antennas to wireless LAN devices. Also, if you are connecting two physically separated LAN sites using wireless LAN equipment, you will most probably need external antennas. You should contact the manufacturer of your wireless LAN equipment for more information on antennas that will be suitable for your needs.

Figure 9.4 A wireless LAN access point.

Networking Support Servers

The networking support servers include the computers that play key roles in a LAN. For example, file servers include large capacity disks and allow file-sharing; network-authentication servers allow network access to only authorized users; and print servers allow sharing of a printer among LAN users. If you are planning to use a network-support server, you should make sure that you have applicable networking hardware available for that server. For example, if you are planning to use a network print server, you should have at least one computer with a printer attached that is connected with the wireless LAN.

Setting Up a Wireless LAN for the 802.11 Infrastructure Mode

The infrastructure mode requires the presence of an 802.11-compliant access point. An access point serves as the backbone of a wireless LAN as all the wireless LAN traffic goes through the AP. In order to use a wireless LAN, you need to configure the APs and all the wireless adapters that will be using the wireless LAN operating in the LAN. A basic wireless LAN operating in infrastructure mode consists of the following components:

Wireless LAN access points. Data in a wireless LAN operating in infrastructure mode is routed among the computers in a wireless LAN via an AP. Today, there are many different flavors of IEEE 802.11-compliant wireless LAN APs available on the market. They differ in the standard they comply with—for example, 802.11a or 802.11b—and the type of additional networking features they may include. For example, some wireless LAN APs come with built-in routers, roaming services, and the Dynamic Host Configuration Protocol (DHCP).

Wireless LAN adapters. Each computer in a wireless LAN must be installed and configured to operate in the infrastructure mode.

It is a good idea to always set up wireless APs before installing computers with wireless LAN adapters. In this section, we help you install a wireless LAN that operates in the IEEE 802.11 infrastructure mode using the Agere ORiNOCO RG-1000 Residential Gateway (AP) and ORiNOCO 802.11b PC Card. Although we target ORiNOCO products when discussing the details of configuration options, we use a general approach that can be applied to most wireless fidelity (Wi-Fi) compliant devices.

Setting Up a Wireless LAN Access Point

Although you do have to carefully choose the physical location at the site where you intend to install a wireless LAN AP, a Wireless LAN AP is the easiest wireless LAN equipment to install. Depending on the operating load of a wireless LAN and/or the physical nature of a deployment site, you may install one or more wireless LAN APs. Following are some of the most common AP installation configurations:

Single AP configuration. The wireless LAN consists of an AP and the wireless workstations associated with it.

Overlapping AP configuration. The wireless LAN consists of two or more adjacent APs whose coverage slightly overlaps.

The multiple AP configurations. The wireless LAN consists of several APs installed in the same location. This creates a common coverage area that increases aggregate throughput.

Many wireless LANs contain several of these configurations at different points in the system. The single AP configuration is the most basic, and the other configurations build upon it. In this section, we guide you through the steps necessary to install a wireless LAN AP using single AP configurations.

Verify the Wireless LAN AP Box Contents

Most APs come with the following items:

Access point. An access point is normally a self-contained box that usually comes with an antenna, a few network jacks on the back, and some status indicator lights in the front.

AC Power adapter. The adapter must conform to the electric outlet voltage available at the site. For example, in the United States, the most common electric power voltage is 110 volts. Using an incorrect voltage adapter could damage the AP and the adapter itself.

Manuals and software disks. Most APs come with a user manual and software disks. If a manual is missing, you should contact the manufacturer or their Web site to obtain the manuals. The software disks may not be present if no software installation is required for configuring the AP. If disks are not present, consult the manual to ensure that you do not need any software to configure the AP.

Wall mount. Some wireless LAN adapters come with a wall mount panel for installation of the AP on a wall. If the mount is present and you wish to install it using the mount, or if the installation requires the use of the mount, you should consult the AP manual for the steps necessary to install the AP using the wall mount.

Ethernet cable. Typically, wireless LAN APs can be connected with a wired LAN or with other networking devices, for example a DSL modem. If your wireless LAN adapter supports such connectivity, then there should be an Ethernet cable. If you do not find an Ethernet cable, you should check with the manual to see whether it is supposed to contain one. You can also purchase this cable from an electronics or computer store.

Write Down the Product Identification Information

Before you proceed, write down the following product information. You may need this information to contact your equipment manufacturer for technical support or warranty purposes:

Manufacturer. The name of the AP's manufacturer. For example, Agere Systems/ORiNOCO.

Model. The product model number as assigned by the manufacturer.

Serial numbers. A serial number uniquely identifying the product.

MAC addresses. MAC addresses uniquely identify a wireless LAN adapter. For more information on MAC addresses, see Chapter 1, "Networking Basics."

Select Access Point Location

The performance of an indoor wireless LAN adapter is greatly affected by the placement of APs. Common factors that affect an AP's performance include physical obstacles, devices operating in the wireless LAN operating frequency, and the presence of heat sources. This section describes various considerations to help you position the AP for optimum coverage and operation of the wireless LAN when performing an indoor, outdoor, or rooftop installation.

Check Electric Power Availability

Make sure that an electric outlet is available or can be provided at the locations where you want to install an AP.

Locate a Central and Highest Location in the Coverage Area for AP

Install the access point at least 1.5 meters above the floor, clear of any high office partitions or tall pieces of furniture in the coverage area. The AP can be placed on a high shelf, or can be attached to the ceiling or a wall using a mounting bracket. Install the AP in a central location in the intended coverage area. Good positions include the center of a large room, center of a corridor, and the intersection of two corridors.

Determine Maximum Possible Distance between an AP and a Wireless LAN Adapter

Wireless LAN APs have a limited operational range. You should consult the AP and the wireless LAN manuals to first find out the maximum distance that you can be between your AP and the wireless LAN adapter. This will ensure that you are optimizing network performance by properly using the full bandwidth of each AP.

198 Chapter 9

Look Out for Competing RF Devices

For best performance, make sure that devices operating within the wireless LAN coverage area do not emit the radio frequency used by the wireless LAN. For example, if you are installing an 802.11b wireless LAN, position the units clear of radiation sources that emit in the 2.4-GHz frequency band, such as microwave ovens.

Heat Sources

Make sure that you select an AP location that is well away from sources of heat, such as radiators and air conditioners.

If you followed these instructions properly, you should have the best locations for your AP. The next step is to install the AP at the location you choose.

Understanding the Agere Systems ORiNOCO RG-1000 Residential Gateway

The Agere Systems ORiNOCO RG-1000 Residential Gateway is an IEEE 802.11b-compliant wireless LAN AP targeted for Home and SoHo users. The RG-1000 Internet Gateway acts as an access point to provide wireless LAN connectivity and Internet access via dial-up connections as well as an Ethernet port. Designed for between one and ten users, the RG-1000 is the ideal Internet access solution for home and small office networks, enabling multiple stations to share a single Internet connection using Network Address Translation (NAT) and Dynamic Host Configuration Protocol (DHCP) functionality (see Chapter 1 for more information on NAT and DHCP). Figure 9.5 shows the ORiNOCO RG-1000 product. Following are some of the key features of RG-1000.

- High performance 11 Mbps data rate
- Wide coverage range of up to 1,750 ft/550 meters
- IEEE 802.11b (Wi-Fi) certified
- Windows 95/98/2000, Me, and Windows NT
- High-level security with 64-bit key WEP encryption and access control table-based authentication

RG-1000 can be used in one of the following network operation modes. You should install and configure the AP to the mode that best suits your needs.

LAN bridge. RG-1000 can be used as a LAN bridge to connect wireless LAN devices to a wired LAN. RG-1000 documentation and configuration software refers to this mode as LAN infrastructure mode. A wireless LAN set up this way relies on the DHCP services of the wired LAN, and the RG-1000 acts only as an intermediary between the computers in the wireless LAN and the wired LAN.

Standalone wireless LAN. RG-1000 can be used to simply interconnect wireless LAN computing devices with each other. A LAN configured this way does not communicate with any outside LAN, and data remains within the wireless LAN.

Internet gateway. RG-1000 can also be configured to act as a wireless LAN gateway to the Internet. When configured as a wireless LAN gateway, RG-1000 does not allow computers in the wireless LAN to communicate with each other but it does allow computers in the LAN to share an Internet connection. RG-1000 supports connectivity to the Internet via phone line or the RJ45 connector that can be used to attach it to a cable, DSL, or ADSL modem.

In the next sections, we continue with the installation process by first installing and configuring the AP, followed by wireless LAN adapter installation and configuration to yield a fully operational wireless LAN.

Install the Access Point

In this section, we install an AP to function in the infrastructure mode. Follow these steps to install the AP:

1. Carefully choose the AP location by following the instruction in the preceding section, *Select Access Point Location*.
2. Connect any network cables that you might need to connect to the AP. For example, if you are connecting the AP with a wired LAN, connect the AP with the wired LAN using an Ethernet LAN cable with the RJ45 network jack normally located on the back of the AP. You must always consult the AP manual before connecting a jack.
3. Power up the AP using the power adapter.
4. When you power up the AP, you should see some lights blinking. If you do not see any such activity, ensure that the power outlet is operational and the power adapter is properly plugged in to both the power outlet and the AP. If you still do not see any light indicators, you should stop here and contact the manufacturer.

Check Antenna Diversity

If your AP uses an external antenna (RG-1000 has internal antennas), make sure the antennas are fully exposed and extended upward vertically in relation to the floor. For models with external antennas, connect the external antennas and RF cable. In most wireless LAN deployments, a single antenna is sufficient to ensure good performance levels. A phenomenon known as multipath propagation, which is caused by reflection of radio waves from potential reflectors, for example automobiles or metal furniture, can degrade wireless LAN performance. In cases where multipath propagation exists, we recommend that an AP with two antennas be used. This takes advantage of space diversity capabilities. By using two antennas per unit, the system can select the best antenna on a per-packet basis. If you are installing on a rooftop or wirelessly connecting two LANs between two buildings or physical sites, you should make sure that the line-of-sight between the antennas is free of any obstacles.

Configuring the Access Point

The access point must be configured before the wireless LAN can become operational. The method for configuring the AP differs from manufacturer to manufacturer. The following are some of the basic methods for configuring the AP.

Wirelessly Configuring an AP Using Vendor-Provided Software or a Web Browser

Most APs are normally shipped in configuration mode. It is often possible to simply power up the AP and use a computer equipped with wireless LAN adapter to configure the AP. The software that is used to configure the AP can be a Web browser, in which case the AP acts as a Web browser and lets you configure the AP parameters, or it could be special software that a vendor provides that helps you with the AP configuration steps. For example, Agere Systems RG-1000 comes with an application program called RG Setup Utility, whereas NetGear MR410 AP (a competitor of RG-1000) can be configured using a Web browser. It is important to remember that in order to wirelessly configure an AP, you must already have at least one wireless LAN adapter installed in one of your computers.

Configuring an AP Using a Wired LAN

An AP can be connected to a wired LAN during configuration if it does not support wireless configuration or if it is to be used as a wireless LAN bridge. Just like configuring an AP wirelessly, an AP can be configured using a Web browser or setup software that is provided by the AP vendor.

Important Wireless LAN AP Configuration Parameters

Most of the wireless LAN AP configuration parameters depend on the type of LAN you are interested in building. These parameters include the wireless LAN Service Set Identifier (SSID) or Network Name, security parameter settings, AP operation mode settings, and AP TCP/IP network settings that include the DNS and IP address settings. Though APs from different manufacturers may use terminologies that may make these settings sound a little different, nonetheless they provide the same basic functions. If you cannot completely understand the AP configuration parameters of the AP you are using, please consult the AP documentation that the manufacturer shipped with the AP or contact the manufacturer. Below is a brief explanation of the basic AP configuration parameters we just mentioned.

- **SSID or Network Name**. Each AP in an IEEE 802.11 wireless LAN must be identified by an SSID or Network Name. The wireless LAN adapters use these SSIDs to identify the APs you want to connect them with. The SSIDs are names, generally up to 32 characters long, that are assigned to each AP. Some APs, for example RG-1000, come with a predefined Network Name, and you must use the given name as SSID or Network Name when configuring the AP or a wireless LAN adapter to work with the AP. Most APs that let you change the SSID do come with a default SSID, but it is not a good idea to use the default SSID, and you should always come up with a new SSID that is easy to remember and makes sense. For example, if you are installing a total of four APs in a four-story building with each floor using one AP, you might want to call the AP on the first floor AP_FLOOR1, the one on the second floor AP_FLOOR2, and so on. Alternatively, you should assign all APs the same SSID if you intend to provide roaming services to the individuals using the wireless LAN equipment. As a wireless LAN adapter travels from the range of one AP into another, it automatically joins the other AP when more than one AP is assigned the same SSID.

- **Security Settings**. 802.11 standard supports Wireless Equivalent Privacy (WEP)-based security. WEP security allows all data communication between wireless LAN clients and/or APs in encrypted form. You should decide whether you want to use the WEP encryption or not. If you decide to use the WEP encryption, you will need to distribute the WEP keys to all wireless LAN adapters. Without WEP keys, wireless LAN adapters will not be able to communicate with the AP. In addition to the WEP security, most wireless LAN APs require a password to administer the AP. Most APs come with a default password, for example the default AP password for RG-1000 is the same as the Network Name as preconfigured by the manufacturer.

- **AP operation mode settings**. As mentioned earlier, you must decide how you want to use your AP. The operation mode of an AP differs from the AP itself. For example, RG-1000 can be used to build a standalone wireless LAN as a wireless LAN gateway to the Internet, or to connect a wireless LAN to a wired LAN. It is important to remember that not all APs provide operating modes like those supported by RG-1000.

TCP/IP Settings. Each computer in a TCP/IP-based wireless LAN must have a unique IP address, must be configured with proper DNS servers, and, optionally, with a default gateway. Client computer and AP configuration will differ for the IP address assignment, DNS settings, and default gateway configurations based on the way the AP will be used. Most APs, for example RG-1000, come with DHCP functionality built-in and can assign IP addresses and other TCP/IP parameters to all wireless LAN adapters that successfully connect with them. If the AP you are using does not support DHCP settings or if you are using the AP only as a bridge between the wireless LAN and the wired LAN, you may have to configure the TCP/IP parameters manually on individual wireless LAN computers and the AP. RG-1000 is capable of assigning TCP/IP configuration parameters to all clients connecting to them. It is also able to obtain TCP/IP information from a DSL or cable modem when connecting a wireless LAN to the Internet.

Since we are using RG-1000 in our example, we will use the software that is provided by Agere Systems. Additionally, we will configure the RG-1000 wirelessly. Since configuring the RG-1000 requires that you must previously have configured a wireless LAN adapter, let's go through the installation and configuration of a wireless LAN adapter and then we will finish configuring the RG-1000.

Setting Up Wireless LAN Adapters

As described earlier, wireless LAN adapters are available for both mobile devices and desktop computers. Typically, PC Card and CF devices are used in the mobile devices, whereas PCI devices are used in the desktop computers. In either case, wireless LAN devices are shipped in sealed boxes containing the wireless LAN adapter in an electrostatic-proof bag, the installation manual, and the software drivers. In this section, we discuss how to install wireless LAN adapters in both mobile devices and desktop computers.

Installing PC Card Wireless LAN Adapter in a Laptop or Notebook Computer

Most new laptops and notebook computers come with PC Card slots. These slots enable you to extend the functionality of your computer by attaching peripherals to it. Many such computers come with two PC Card slots. You will need to use one of these slots to use the wireless LAN adapter in your computer. Normally, PC Card slots are located on the sides of a notebook or laptop computer as shown in Figure 9.5.

Follow the steps below when installing a PC Card wireless LAN adapter in a notebook computer with PC Card slot.

1. Locate an empty PC Card slot (a slot that does not have a PC Card device already installed). As mentioned earlier, these slots are located on the side of most notebook computers. If you cannot locate them, you should contact your notebook manufacturer. If you successfully locate them, but you find that all available slots (each laptop normally contains one or two PC Card slots) are already taken by other devices, for example a modem or a wired LAN adapter, you will have to remove one of the adapters and replace it with the wireless LAN adapter. If you are successful in locating an empty PC Card slot, you are in luck and you should move on to the next step; otherwise, you might have to reconsider using a wireless LAN adapter with your computer.

2. Power down and disconnect any power cables to your notebook computer before you place the wireless LAN adapter in your computer.

3. PC Cards normally have direction signs printed on them (as arrows). You should insert the PC Card with the side that has the direction printed on it aligned with the top of the computer. The process of inserting a PC Card wireless LAN adapter is shown in Figure 9.5.

Figure 9.5 Installing a PC card wireless LAN adapter in a notebook computer.

Installing a CF Wireless LAN Adapter in a PDA

The steps involving the installation of a CF wireless LAN adapter in a PDA are very similar to those of installing a PC Card. The process normally involves sliding the CF adapter into a PDA in correct alignment, with the PDA powered down.

If you are interested in using a CF card with a notebook computer, you will need to purchase a CF jacket or CF adapter for PC Card card. CF jackets are available at computer hardware stores and look almost like the PC Card cards. To install the CF-based adapter, first insert the CF adapter into the PC Card adapter, and then install the PC Card adapter as described earlier in this chapter.

Installing a PCI Wireless LAN Adapter in a Desktop Computer

Adding a wireless LAN adapter to a desktop computer involves more work than installing an adapter to a PDA or notebook computer. The following are the steps involved when installing a PCI wireless LAN adapter to a desktop computer.

1. Power down and disconnect any power cables to your desktop computer before you install the wireless LAN adapter in your computer.

2. Open the desktop computer according to the manufacturer's instructions. Locate an empty PCI slot (a slot that does not have a PCI device already installed). Generally speaking, PCI slots are white slots normally located on the left-hand side of the computer motherboard (the main electronic circuitboard that most of the electronic circuitry is soldered to). If you do not have an empty PCI slot, you will either have to remove one of the cards, sacrificing the functionality that the card provided you, or you will not be able to install the wireless LAN adapter on your desktop computer. If you choose to remove one of the cards, be careful which one you remove. Removing a card could result in either a lack of functionality or in your computer becoming inoperable due to the absence of the card you removed. If you are successful in locating an empty PCI slot, you are in luck and you should move on to the next step; otherwise, you might have to reconsider using wireless LAN adapter with your computer.

3. PCI standard adapters have notches, which let them install only in the correct direction. Also, most PCI adapters are attached to a metal plate that faces the outside of the computer. You should carefully insert the PCI adapter into the slot.

If the slot available in your computer seems bigger or smaller than the card, do not try to install the card into the slot. Contact the manufacturer of the computer to find out if you really have a PCI slot in your computer. In cases where the PCI card does not fit the available slot, you might have slots that comply to other standards. Contacting the manufacturer of your computer might help you locate the right type of wireless LAN adapter that will work with your computer.

Introducing the ORiNOCO 802.11b PC Card

ORiNOCO 802.11b PC Card is an IEEE 802.11b standard-compliant wireless LAN adapter that can be used in computers, normally PDAs and laptop computers, that have either a PCMCIA II or a CardBus slot. The ORiNOCO PC card can be used anywhere to connect to a wireless Ethernet network and is interoperable with Wi-Fi-compliant products. The card delivers high-speed wireless networking at 11 Mbps, operating in the 2.4 GHz unlicensed frequency. The ORiNOCO PC Card is shown in Figure 9.6.

The following are some of the key features of ORiNOCO PC Card.

- Plugs directly into laptop type-II PCMCIA slot
- IEEE 802.11b (Wi-Fi) certified
- Low power consumption
- High performance 11 Mbps data rate
- High-level security with full 128-bit key or 64-bit WEP encryption
- Operating systems supported: Novell Client 3.x and 4.x, Windows 95/98/2000, Me, and Windows NT (NDIS Miniport driver); Apple Macintosh; Windows CE; and Linux

Figure 9.6 Agere Systems ORiNOCO PC Card.

Installing and Configuring Wireless LAN Adapter Software

In order to use the wireless LAN equipment in a LAN, computers with wireless LAN adapters must be installed with proper software. This software is normally shipped by the manufacturer or can be obtained from the manufacturer's Web site. Though the basic configuration parameters for all wireless LAN adapters are the same, the actual configuration of wireless LAN adapters is operation-mode dependent.

Most wireless LAN adapters come with software drivers suitable for popular operating systems. You must install the vendor-provided software to be able to use the wireless LAN. Before proceeding further, if necessary, make sure that all applicable network support components (for example, access points, DHCP servers, and authentication servers) are fully configured and operational.

Follow these steps to ensure proper software setup on the computer that you want to connect with a wireless LAN. The steps that follow assume that you have already installed the physical wireless LAN adapter in the computer:

1. Locate the appropriate software for the operating system of your computing device. For example, if you want to use an ORiNOCO wireless LAN adapter with your notebook installed with Microsoft Windows XP, you should look for driver files for Windows XP in the compact disk (CD) that Agere Systems ships with the adapter. If you cannot find the appropriate driver, in this case for Windows XP, you should contact the manufacturer if it supports Windows XP. If a manufacturer does not support the OS you are using, you might not be able to use that wireless LAN adapter with that computer. In such a case, you might want to exchange the wireless LAN adapter for one that supports the OS of your choice.

2. Turn on your computer. If your computer supports multiple users, log in to your computer as a user who has privileges that allow him or her to install software on the computer. Depending on the operating system, this privileged user is normally known as the administrator or root user. If you are using a plug-and-play operating system, such as Windows 98, Windows ME, Windows 2000, or Windows XP, the system will automatically recognize that a new device has been installed and will prompt you for software installation and/or configuration.

In this section, we walk you through the steps involved in installing the wireless LAN adapter software for the ORiNOCO Silver PC Card under Microsoft XP. For additional example installations, please see Appendix B.

Setting Up ORiNOCO PC Card under Microsoft Windows XP

Microsoft Windows XP has built-in support for most popular wireless LAN adapters. The ORiNOCO PC Card drivers are included with the Windows XP Operating System. The original Windows XP version 5.1, build 2600, contains ORiNOCO driver version 7.14. Assuming your hardware is functioning properly, Windows XP will automatically load this driver. Follow these steps to configure Windows XP for using the ORiNOCO Silver PC Card.

1. If you have not already inserted the PC Card into your laptop, insert it carefully using the manufacturer's instructions.

2. Windows XP will immediately prompt you with a message, Found New Hardware, in your system tray as shown in Figure 9.7. If this dialog does not appear, consult the ORiNOCO user's guide for more information on Windows XP support for your wireless LAN adapter.

3. Using the right button of your mouse, click the network icon in the system tray. A Network Connections menu, similar to that shown in Figure 9.8, will appear.

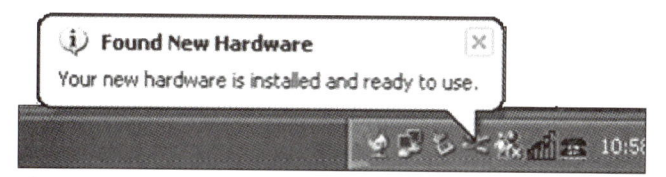

Figure 9.7 New hardware detection dialog under Windows XP.

Figure 9.8 Network Connections menu under Windows XP.

Chapter 9

4. Select the View Available Wireless Networks menu item. Connect to Wireless Networks dialog will appear. Click Advanced in this dialog.

5. The Wireless Network Connection Properties dialog will appear as shown in Figure 9.9.

6. You should see the name of your RG-1000 AP under Available Networks. If you are using more than one AP, they all should be listed. If you cannot find the AP you just installed, you should make sure that you correctly installed the AP. In addition to the AP being listed under the list of Available Networks, you will also see green lights blinking on the AP. Consult the AP manual for more information on AP led indicators.

7. If your Network Name does appear under Available Networks, select your Network Name and click Configure. The Wireless Network Properties dialog appears as shown in Figure 9.10.

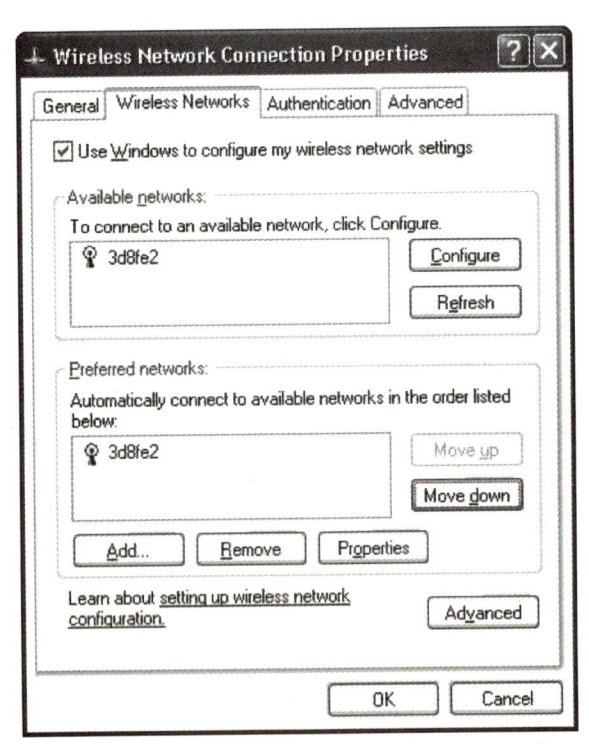

Figure 9.9 Wireless Network Connection Properties screen.

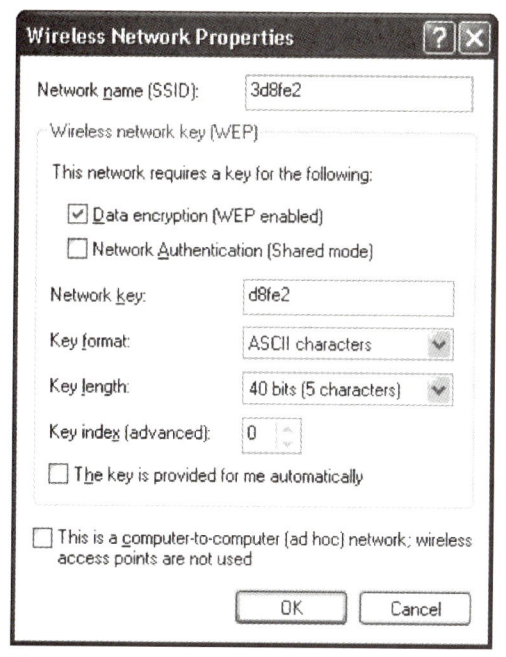

Figure 9.10 Wireless Network Properties screen.

8. ORiNOCO RG-1000 comes with security enabled by default; you will have to enable security to successfully connect to the ORiNOCO RG-1000. Unless modified by Agere or someone who previously altered the default settings of the RG-1000, the default SSID of the RG-1000 is the last five characters of the AP Network Name. For example, if the AP Network Name was 3df8e2, the default encryption key is df8e2. Check the Data Encryption check box and type the encryption key in the Network key field. Close the Wireless Network Properties screen by clicking OK. Also dismiss the Wireless Network Connection Properties dialog by clicking OK.

You are now finished with the wireless LAN adapter and network configuration for Windows XP. Now that you are done with installing wireless LAN adapter software, in the next section we talk about configuring the wireless LAN AP equipment to form an operational wireless LAN.

Finishing the Access Point Configuration

Finishing the AP configuration step involves finalizing the AP configuration settings that yield an operational wireless LAN. In this section, we configure the RG-1000 for standalone wireless LAN operation. The wireless LAN resulting from this configuration allows computers in the wireless LAN to communicate with each other, as shown in Figure 9.11. In a later section, *Connecting Wireless LAN Based on ORiNOCO RG-1000 to the Internet,* we explore the other operating modes of RG-1000 that you can use to connect a wireless LAN to the Internet.

In this section, we assume that you have installed and powered up the RG-1000 AP and successfully installed the ORiNOCO Silver PC Card wireless LAN adapter. Follow the steps in the next section to successfully install RG-1000 Setup Utility software and configure the RG-1000 AP for standalone wireless LAN operation.

Install the ORiNOCO RG-1000 Setup Utility Software

ORiNOCO ships a compact disk (CD) with the RG-1000 AP that contains the RG Setup Utility. This is the software that you need to wirelessly configure the RG-1000. Following the steps will help you successfully install the RG Setup Utility.

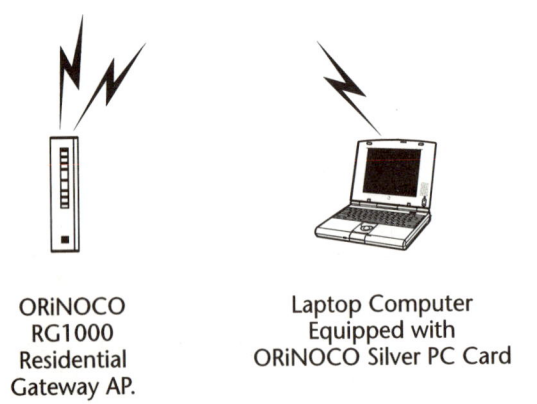

Figure 9.11 Standalone wireless LAN using ORiNOCO RG-1000 and ORiNOCO Silver PC Card.

Equipment Provisioning and LAN Setup 211

1. Insert the CD-ROM that came with your Residential Gateway kit into your computer. Your operating system will run the CD automatically.
2. Click the install button RG Setup Utility.
3. Select Install Software.
4. Select Install Client Manager. Follow the instructions on your screen.

If the CD-ROM does not start automatically, do the following:

5. Click the Windows Start button.
6. Select Run.
7. Browse to the CD-ROM.
8. Double-click the file "start.exe".

Once the installation process is finished, you can execute the RG Setup Utility to set up the RG-1000 AP.

Start the RG Setup Utility

The RG Setup Utility allows you to view or modify the following network settings.

Internet Access Settings. The Internet access settings enable you to use the RG-1000 gateway to share an Internet connection or to connect the wireless LAN to a wired LAN.

Wireless Connection Settings. Wireless connection settings for radio channel, security, and TCP/IP parameters of the wireless LAN.

The following steps help you start the RG Setup Utility:

1. Click the Start button on the Windows task bar.
2. Select Programs, and then select ORiNOCO.
3. Select RG Setup Utility to start the program.
4. Once the RG Setup Utility starts executing, you see a Welcome screen similar to Figure 9.12.

Click the Continue button. You should see a screen similar to one shown in Figure 9.13. To connect to RG-1000, enter the six-character Network Name printed on the label on the device and click Continue.

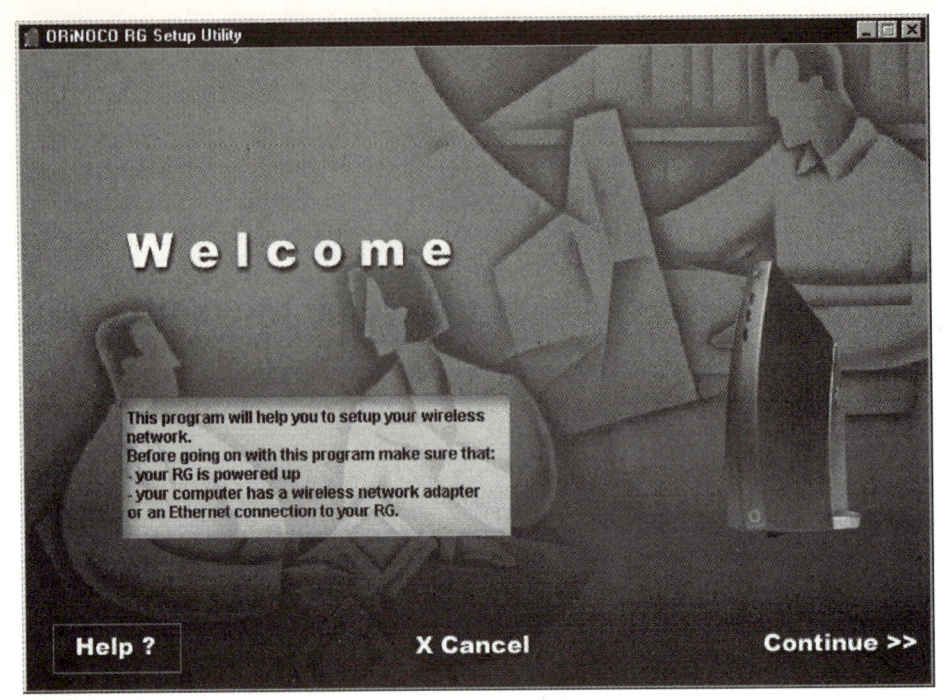

Figure 9.12 ORiNOCO RG-1000 (Residential Gateway) RG Setup Utility.

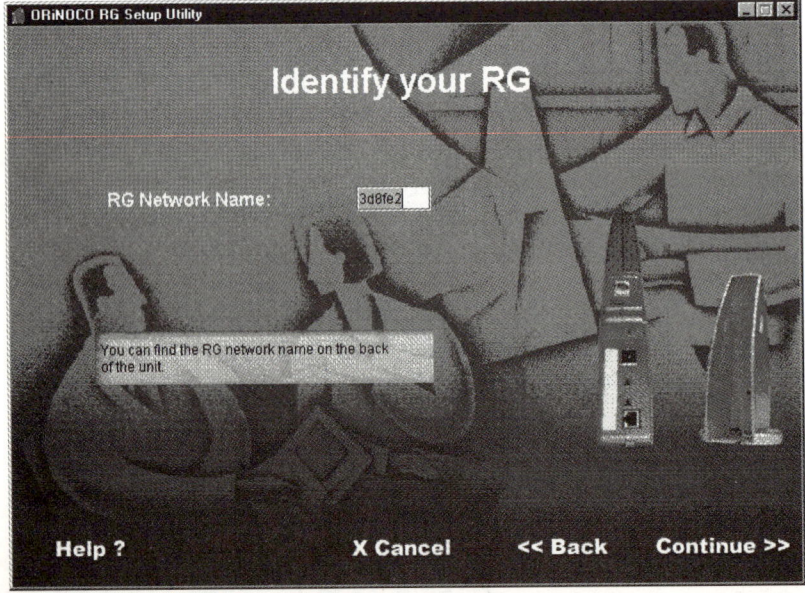

Figure 9.13 Connecting to RG-1000 using RG Setup Utility.

Equipment Provisioning and LAN Setup 213

If a screen Setup your Internet Connection, as shown in Figure 9.14 appears, you were successful in properly starting the RG Setup Utility. If you encounter difficulty accessing the Residential Gateway to view or modify its current settings:

1. View and modify, if needed, the settings of the wireless adapter in your computer to ensure that the Network Name matches the value printed on the label at the back of the Residential Gateway.

2. Make sure that the encryption key matches the value of the Residential Gateway (default key matches the last five digits of the Network Name).

3. View and modify, if needed, the Networking properties on your computer to ensure the TCP/IP protocol is installed for your wireless adapter. Make sure that the TCP/IP protocol has DHCP enabled, to obtain an IP Address from the Residential Gateway automatically.

Modify the Internet Access Settings

The Setup your Internet Connection screen lets you decide how you want to use your RG-1000. Since, in this example, we only want to set up a standalone wireless LAN, select [no Internet] option for Internet access via option. Figure 9.14 shows the selection of this option.

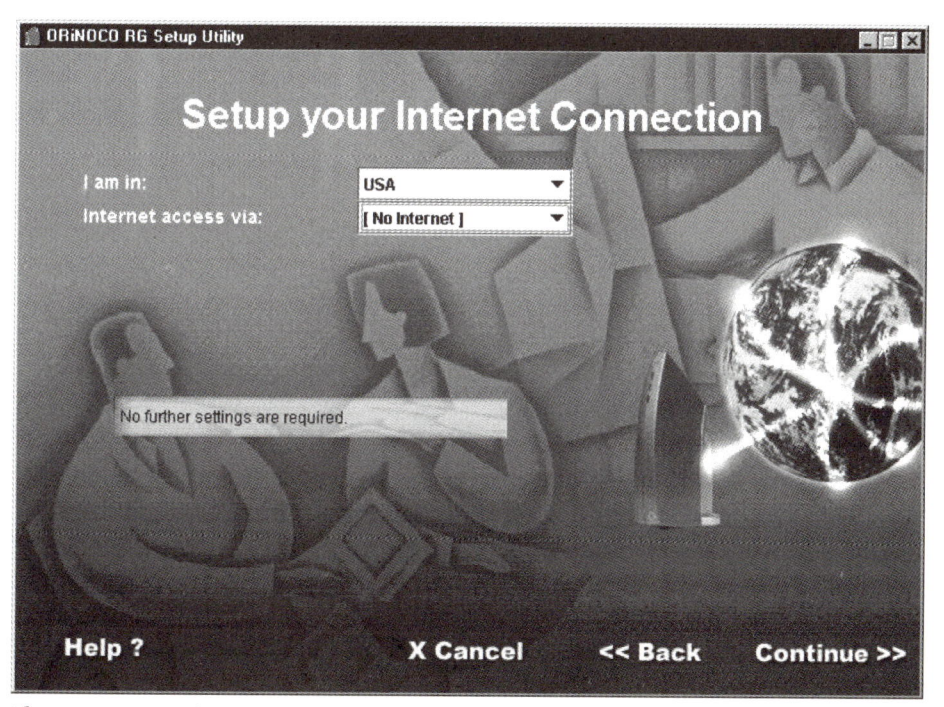

Figure 9.14 Setting up Internet connection options for RG-1000.

Chapter 9

Modify Residential Gateway Wireless Connection Settings

Change these settings to increase the security of your wireless network, set up special connection requirements, and improve your wireless communication. Figure 9.15 shows the wireless connection settings.

Wireless channel. To transmit and receive data, the Residential Gateway uses a frequency channel. If neighboring wireless networks are using the same channel, it is advisable to have your Residential Gateway network use a different channel.

Encryption key. Communication within your network is only possible by wireless computers using the same Encryption Key. This is what the Residential Gateway uses to enable Data Security on your wireless network. The default value of the Encryption Key equals the last five characters of the Network Name. To prevent any access to your network without permission, it is strongly advised to change the default encryption key value.

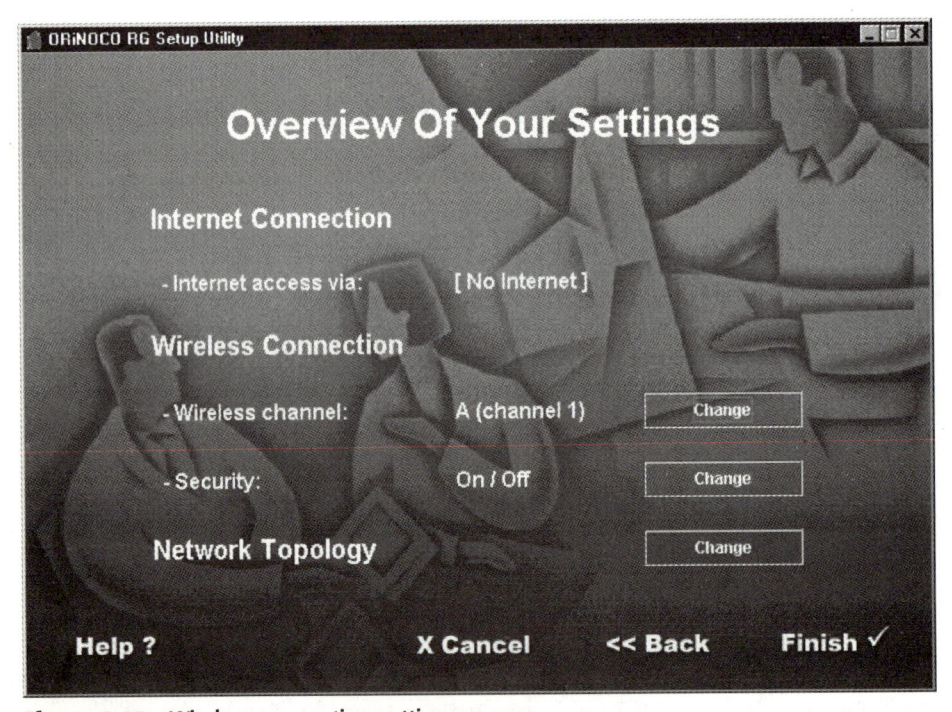

Figure 9.15 Wireless connection settings screen.

Click on Finish to complete the AP setup process. If you do not get an error message when you click on Finish, you have successfully completed the RG-1000 setup process. If you get an error message, make sure that you have correctly followed the configuration steps and consult the ORiNOCO RG-1000 user's manual for more information.

Once you have completed the RG-1000 setup process and successfully installed the ORiNOCO PC Card in your laptop, you have completed the steps required for your standalone wireless LAN. The next step for you is to test the wireless LAN from the laptop computer and add any other computers to the Internet. If you want to connect the RG-1000 to the Internet, see the section *Connecting the Wireless LAN to the Internet Using ORiNOCO RG-1000* later in this chapter.

Testing Your Standalone Wireless LAN

Follow the next steps to test the standalone wireless LAN that you just built using the ORiNOCO RG-1000 Gateway and the ORiNOCO Silver PC Card.

1. Connect to your wireless network from Windows XP computer by selecting View Available Wireless Networks from the menu as shown in Figure 9.8. The Connect to Wireless Network dialog box appears. Select your wireless LAN AP from the list of available networks and click on Connect. You should now be connected to the wireless LAN.

2. On your Windows XP computer, start a DOS window by clicking on Start menu and then clicking on the Run menu item. If you are using Windows XP Professional, type **CMD**; if you are using Windows XP Home Edition, type **command**.

3. From the DOS window, type the following:

   ```
   C:>ping 10.0.1.1
   ```

If you get a response as shown below, you have correctly installed and configured your wireless LAN:

```
Pinging 10.0.1.1 with 32 bytes of data:

Reply from 10.0.1.1: bytes=32 time<10ms TTL=128
Reply from 10.0.1.1: bytes=32 time<10ms TTL=128
Reply from 10.0.1.1: bytes=32 time<10ms TTL=128
Reply from 10.0.1.1: bytes=32 time<10ms TTL=128
```

If there is an error or the computer is not configured correctly, you might get an error as follows:

```
C:\>ping 10.0.1.1
Pinging 10.0.1.1 with 32 bytes of data:
Request timed out.
Request timed out.
Request timed out.
Request timed out.
```

If you get the error message, check the operating system instructions and the instructions provided by the network card provider and by this chapter to make sure that you have followed the instructions correctly.

Adding More Computers to Your Standalone Wireless LAN

In order to add more computing devices to your wireless LAN, add the wireless LAN adapters and configure them according to the instructions provided in the section *Setting up Wireless LAN Adapters* earlier in this chapter. You will not have to make any changes to your RG-1000 unless you need to change the TCP/IP settings or you want to add more APs to the wireless LAN.

Connecting a Wireless LAN to the Internet

Today, one of the most popular uses of a wireless LAN is to share an Internet connection among its users. There are many different methods by which a wireless LAN can be connected to the Internet. The following are some of the more common:

Directly connecting the access point to a broadband connection. Some APs, for example ORiNOCO RG-1000, come with the capability to communicate with a DSL or cable modem to share an Internet connection among its users.

Connecting the access point with a wired LAN that is already connected to the Internet. Most APs, for example ORiNOCO AP-1000, are able to connect the wireless LAN with a wired LAN. If an Internet connection is already available to the wired LAN and the routers are configured such that traffic between the wireless LAN and the Internet, for example, a broadband DSL connection, is allowed, the network users in the wireless LAN can easily communicate with the Internet.

It is important to remember that connecting a wireless or wired LAN to the Internet opens up your LAN to users on the Internet and your network becomes vulnerable to attacks from hackers. We strongly recommend that when planning to connect your computer to the Internet, you should make sure that you use, at least, a firewall and a router to control the traffic between the Internet and your LAN. You should also make sure that you take necessary steps to ensure the continued security of your LAN. These steps normally include keeping in touch with the computer networking community and the networking industry to keep you aware of new developments in security standards and preventive measures against known and newly discovered security threats.

In this section, we discuss how wireless LANs that are built using the ORiNOCO RG-1000 Residential Gateway can be connected to the Internet.

Connecting Wireless LAN Based on ORiNOCO RG-1000 to the Internet.

The ORiNOCO RG-1000 can be connected to the Internet by configuring the Setup your Internet Connection screen using the RG Setup Utility as shown in Figure 9.16.

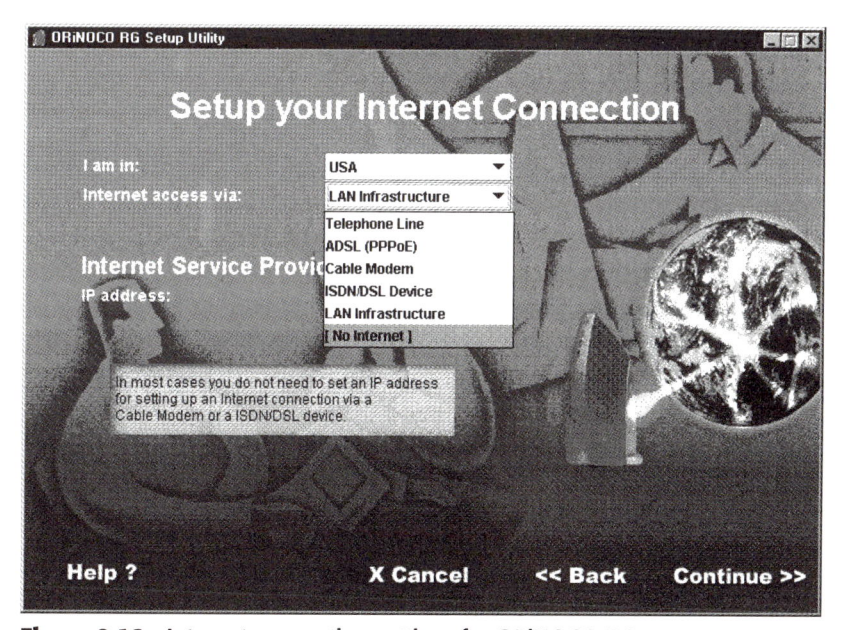

Figure 9.16 Internet connection settings for ORiNOCO RG-1000.

RG-1000 provides four different ways to connect a wireless LAN to the Internet:

- Connecting a wireless LAN to a wired LAN that is already connected to the Internet. Select the [LAN infrastructure] option when choosing to connect with the Internet using this method. You will need to make sure that you have all routers properly configured to allow network traffic to flow between the wireless LAN and the Internet through the wired LAN.

- Choose the [Telephone Line] option to connect the wireless LAN to the Internet using the built-in V-90 modem. You will need a dial-up Internet account and a phone line to be able to use this option. Contact your Internet Service Provider (ISP) for more information on how you can use this option.

- Use the [ISDN/DSL Device] or [ADSL] Option to connect with the Internet using a DSL, ISDN, or ADSL connection. You will need a respective technology's modem and an account with the service provider.

- Similarly, for using cable-based Internet connection, you will need a cable modem and a cable Internet account with your cable provider.

Using Multiple AP Configurations

Wireless LANs that have higher bandwidth needs, that serve more users than a single AP can serve, or that need to provide coverage over an area that cannot be covered by a single AP, can use additional APs in the geographical coverage area to provide high bandwidth, more load, and greater coverage area. The following are some of the most popular ways to use more than one AP to enhance the performance of a wireless LAN.

Overlapping AP Configuration

When adjacent access points are positioned close enough to each other, parts of the coverage area of access points may overlap with each other. For example, as shown in Figure 9.17, when two adjacent access points, A and B, are positioned close enough to each other, a part of the coverage area of access point A overlaps that of access point B.

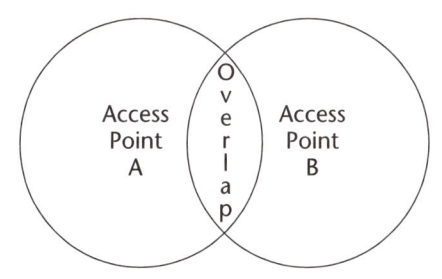

Figure 9.17 Overlapping access points.

This overlapping area has two very important attributes.

- Any workstation situated in the overlapping area can associate and communicate with either AP A or AP B.
- Any workstation can move seamlessly through the overlapping coverage areas without losing its network connection. This attribute is called seamless roaming.
- Follow these steps to properly configure two APs in an overlapping configuration.

    1. Install an AP using the instructions described earlier. Be sure to position the AP at the highest point possible.
    2. Install the second AP so that the two are positioned closer together than the prescribed distance according to the manufacturer's instructions.
    3. To allow roaming, configure all APs and station adapters to the same SSID.
    4. To improve collocation and performance, configure all APs to different hopping sequences of the same hopping set.
    5. Position the wireless workstation at approximately equal distances from the two APs.
    6. Temporarily disconnect the first AP from the power supply. Verify radio signal reception from the first AP using a wireless LAN computer equipped with a wireless LAN adapter.

7. Disconnect the second AP from the power supply and reconnect the first AP. Again, verify radio signal reception from the second AP using a wireless LAN computer equipped with a wireless LAN adapter.
8. If necessary, adjust the distance between the APs so that the coverage areas overlap.
9. Continue setting up overlapping cells until the required area is covered.

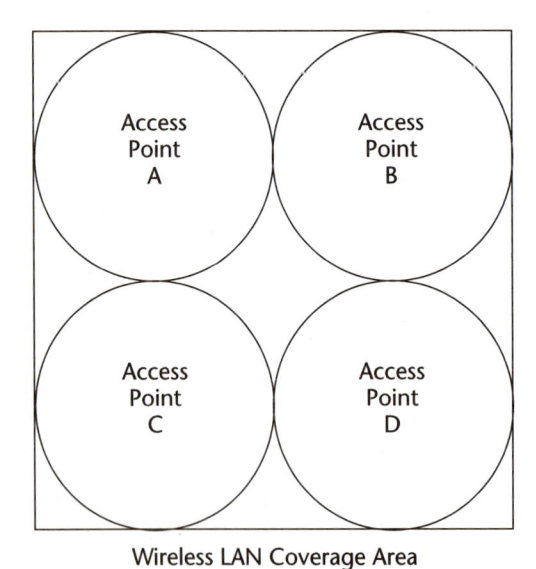

Non-Overlapping AP Configuration

Wireless LANs congested with many network users and heavy traffic load may require non-overlapping, multi-AP configuration. In a non-overlapping, multi-AP configuration, several APs are installed in the same location. The non-overlapping, multiple AP configuration is shown in Figure 9.18.

Wireless LAN Coverage Area

Figure 9.18 Non-overlapping AP configuration.

In non-overlapping configuration, each AP has the same coverage area, thereby creating a common coverage area that increases aggregate throughput. Any workstation in the overlapping area can associate and communicate with any AP covering that area. Follow these steps to set up a wireless LAN with multiple APs in a non-overlapping configuration.

1. Calculate the required number of APs as follows: Multiply the number of active users by the required throughput per user, and divide the result by 1.5 Mbps (which is the net throughput supported by collocated APs). Consider the example of five active stations, each requiring 0.5 Mbps throughput. The calculation is (5*.5)/1.5=1.6. Two APs should be used. This method is accurate only for the first few APs.

2. The aggregate throughput of the common coverage area is equal to the number of colocated APs multiplied by the throughput of each individual AP, minus a certain amount of degradation caused by the interference among the different APs.

3. Install several APs in the same location a few meters from each other so they cover the same area. Be sure to position the APs at the highest points possible.

4. To allow roaming and redundancy, configure all APs and station adapters to the same SSID.

5. To improve collocation and performance, configure all APs to different hopping sequences of the same hopping set.

6. Install wireless LAN adapter cards on computers that need to be connected to the wireless LAN.

It is important to note that dead spots are more prevalent in the non-overlapping configurations as there can be areas in a deployment that do not get covered. However, in closed areas, the dead spots are often eliminated by the radio wave reflections from walls and other objects. In open areas, placing APs in regions that lack coverage can eliminate the dead spots. For example, as shown earlier in Figure 9.16, the area between the four APs is not covered by any AP and is a dead spot. Placing a fifth AP in the middle can easily cover this region. The fifth will partially overlap the other four APs and will eliminate the dead-spot problem. This combination of non-overlapping and overlapping APs is shown in Figure 9.19.

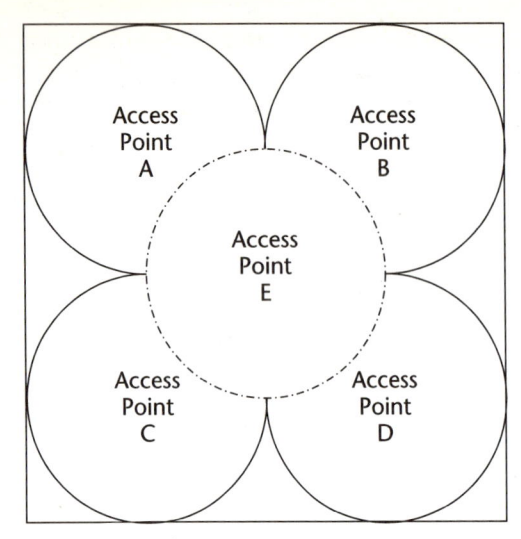

```
                    ┌─────────────────────┐
                    │   Access     Access │
                    │   Point      Point  │
                    │     A          B    │
                    │                     │
                    │       Access        │
                    │       Point         │
                    │         E           │
                    │                     │
                    │   Access     Access │
                    │   Point      Point  │
                    │     C          D    │
                    └─────────────────────┘
                      Wireless LAN Coverage Area
```

Figure 9.19 Eliminating dead spots by using an overlapping AP.

Setting Up Wireless LAN for the 802.11 Ad-Hoc Mode

802.11 ad-hoc mode enables the wireless LAN devices to communicate directly with one other. Ad-hoc LAN is also known as peer-to-peer wireless LAN. For example, two notebook computers installed with a wireless LAN adapter can communicate with each other, forming an ad-hoc LAN, if they are configured for ad-hoc mode operation. When comparing 802.11 ad-hoc mode with the 802.11 infrastructure mode, the most significant different is absence of an AP in ad-hoc mode, whereas you must use an AP in the infrastructure mode. Ad-hoc mode is only used in situations where a small number of computers are to be interconnected for short periods of time and, in this section, we only briefly talk about the steps necessary to set up an ad-hoc wireless LAN. For more information on setting up an ad-hoc wireless LAN, consult your wireless LAN adapter user's guide. Following are the basic steps for configuring a wireless LAN adapter to operate in the ad-hoc mode.

1. Install wireless LAN adapter and software on each computer as described in the section *Setting up Wireless LAN Adapters*.

2. Make sure that wireless LAN adapters in all computers are configured to operate in ad-hoc mode.

3. Configure the TCP/IP settings. Since the ad-hoc mode computers communicate with each other directly, you will most probably have to configure the TCP/IP parameters for each computer in the ad-hoc wireless LAN.

Summary

Though wireless LANs are simple to install, the variety of configuration operations can be make the installation process quite complicated. In this chapter, we explored the process of setting up a wireless LAN that operates in infrastructure mode using the ORiNOCO Residential Gateway as AP and ORiNOCO PC Card. We discussed the installation and configuration of AP and the wireless LAN adapter. We also explained how you could connect a wireless LAN to the Internet using various LAN configurations. In the end, we discussed the commonly used AP configurations to enhance the performance of a wireless LAN.

In this chapter, we built a wireless LAN that can be connected to the Internet and that uses the default 802.11 standard security protocol. In the next chapter, "Advanced 802.11 Wireless LANs," we explore the process of building wireless LANs that provide high security using the 802.1X protocol and the virtual private network (VPN) technology.

CHAPTER 10

Advanced 802.11 Wireless LANs

Wireless LANs based on basic 802.11 standard technologies, for example 802.11b and 802.11a, provide a level of security that is usually sufficient for home, SoHo, small enterprise, or WISP needs. For large enterprise and environments where a fault-tolerant and disaster-resistant LAN is desired, the basic 802.11 equipment alone might not be sufficient. Most large deployments have requirements that basic 802.11 either cannot fulfill or would require supplemental technologies. These requirements normally include high security and authentications, point-to-point connectivity, and secure remote access.

In this chapter, we describe using the supplementary LAN technologies and hardware that can be used to solve authentication, point-to-point connectivity, and remote access over the Internet using virtual private network (VPN) requirements.

High Security and Authentication–Enabled 802.11 Wireless LANs

The IEEE 802.11b standard suggests wired equivalent privacy (WEP) protocol-based security. WEP is considered inherently insecure due to the weaknesses of the encryption keys it uses (for more information on WEP security, see Chapter 7, "Planning Wireless LANs"). WEP security lacks the following two important features:

Authentication. WEP only authenticates the physical hardware that a LAN client might be using. For example, a wireless LAN that uses WEP as the only authentication mechanism does not know who is the actual user using the authenticated wireless LAN adapter. Therefore, stolen laptops or forged MAC (media access control) addresses can be used to infiltrate the network. For any serious wireless LAN deployment, all users using wireless LAN equipment must be authenticated.

High data privacy. The WEP encryption protocol, as defined in the 802.11b standard, is highly criticized and has been proven insecure. The strong encryption in a wireless LAN is extremely important where the data is considered private and confidential.

Currently the 802.1X protocol is used to solve the authentication problem, and using VPN technology solves the data privacy issue. In this section, we first briefly discuss the 802.1X standard (see Chapter 6, "Securing the IEEE 802.11 Wireless LANs," for detailed information on 802.1X) and the VPNs; then we walk through the steps to build a secure 802.11-based wireless LAN that uses these two technologies.

The 802.1X Standard

The 802.1X authentication protocol was originally designed for wired Ethernet-based LANs, but it can also be applied to 802.11-based wireless networks. The 802.1X standard provides user authentication instead of hardware authentication. Besides the authentication features, 802.1X also contains a built-in fast rekeying of the WEP keys, which makes WEP a bit more secure. To use 802.1X in an 802.11 wireless LAN, you must use the following components:

- 802.11a- or 802.11b-compliant wireless LAN adapter(s).
- Software drivers for the wireless LAN adapters, which support the 802.1X protocol.
- 802.11a or 802.11b wireless LAN APs that support 802.1X protocol.
- A software– or hardware-based authentication server, for example a Remote Authentication Dial-in User Service (RADIUS) Server, can be used as an authentication server (see Chapter 6 for more information on RADIUS servers).

Note that if you already have an 802.11-based wireless LAN, you might be able to use the existing 802.11b and 802.11a wireless LAN adapters provided you can obtain the 802.1X-enabled software drivers for them. You also might be able to use the existing hardware with new firmware from the manufacturer. However, you will very likely have to purchase new 802.1X-enabled APs. Contact your hardware manufacturer to find out the additional equipment you will need to build a wireless LAN that uses the 802.1X authentication protocol. For more information on 802.1X, see Chapter 6, "Securing the IEEE 802.11 Wireless LANs."

Virtual Private Network for Wireless LANs

Virtual private networks (VPNs) provide location transparency (users in a VPN enjoy seamless network services without much hint that they are not physically connected to a network) by routing the IP packets, and they provide data security over an insecure medium by transmitting data over the transmission link in an encrypted form. VPNs are normally used in the following situations.

Connecting a remote computer to a private LAN over an insecure medium. Most businesses these days allow their workers to work from home or remote sites. These employees often need access to the corporate data and computing resources. VPNs provide this service by allowing a remote user to securely connect to a LAN over the Internet.

Connecting two LANs over an untrusted or insecure medium. VPN technology can also be used to interconnect physically separate private LANs over the Internet, or to provide secure communication among devices requiring high security in an untrusted or semitrusted LAN. Examples of untrusted LANs include a private LAN, which allows users to use the LAN without requiring authentication. A semitrusted LAN is one that requires authentication, but does allow outsiders to use the LAN.

Most VPN solutions consist of two basic components: the VPN client and the VPN gateway. User computers are usually equipped with the client components, whereas the LAN to which users connect using VPN are equipped with the gateway.

VPN Gateways

VPN gateways are normally installed at the VPN site to which a user or a remote site intends to connect. VPN gateways are normally installed outside the demilitarized zone (between the communication provider equipment and the firewall). VPN gateways facilitate the VPN connectivity between a protected LAN and a remote VPN peer (a VPN client or gateway) by acting as a broker between the two entities and allowing data from only authenticated users to reach the private LAN and vice versa, hence providing a virtual LAN connectivity to the remote peer. Since the data transmission in VPN is always encrypted, hackers cannot tamper with the data or gain access to the remote LAN. VPN gateways are generally required to be high-performance network devices as more than one computer may connect with them at one time. VPN gateway performs services to establish a VPN connection, which are explained next.

Authentication

VPN gateways generally use a network operating system user database, an LDAP (lightweight directory access protocol) directory, or a separate authentication server to authenticate the users authorized to use the VPN connectivity. In addition to the username and password-based authentication, a VPN gateway may also use time-synchronous tokens, for example RSA SecureID, for authentication purposes. For more information on RSA Security, visit www.rsasecurity.com.

Data Privacy through Encryption

VPN gateways use cryptographic encryption algorithms and protocols to provide data security. The most commonly used protocol is known as Internet Protocol Security (IPSec), and the most commonly used encryption algorithm is known as Triple-Digital Encryption Standard (Triple-DES or 3-DES).

Dynamic Host Configuration Protocol (DHCP) and Network Address Translation (NAT) Services

VPN gateways act as a Dynamic Host Configuration Protocol (DHCP) server and assign each VPN peer (a client or another gateway) a unique IP address that does not belong to the protected LAN. When data is received from the VPN peer for the protected LAN or from the protected LAN for the VPN peer, VPN gateway performs the translation of the addresses and transmits the data to the intended party. For example, let's assume that, upon successful authentication, a VPN gateway assigns an IP address 192.168.0.10 to a VPN peer, and the LAN that the VPN gateway was protecting uses 100 IP addresses from 193.168.1.100 to 193.168.1.200. In this case, the VPN gateway may create an entry in a table, called a network address table, that consists of two IP addresses, one that was assigned to the VPN peer and the other an unused IP address from the protected LAN. This entry could look like the one shown in Table 10.1.

Table 10.1 A Sample Network Address Table with One Entry

PEER IP ADDRESS	LAN IP ADDRESS
192.168.0.10	193.168.1.201

When the VPN gateway receives data from the VPN peer, it performs a network address table lookup and an address translation (substitutes the address in the data packet from 192.168.0.10 to 193.168.1.201) so that the data packet can be recognized and properly delivered in the protected LAN. The VPN gateway performs a reverse translation when data originate from a protected LAN intended for the VPN peer. This translation of the IP address is known as Network Address Translation (NAT).

VPN gateways authenticate users, provide data privacy, and act as routing agents by assigning virtual IP addresses (IP addresses that are not part of the LAN) to the VPN clients and translating them to real addresses.

VPN Clients

A VPN user's computer is normally equipped with a VPN client. The VPN client software facilitates VPN connectivity between a VPN gateway and the user's computer by providing the authentication information to the VPN gateway, obtaining and assuming the IP address from the VPN gateway, and performing encryption and decryption operations on all TCP/IP data transmission between the client computer and the VPN gateway. For a VPN client to successfully establish and maintain a connection, it must use encryption algorithms, authentication, and VPN protocols that are compatible with the VPN gateway.

Depending on the deployment nature, security, and performance requirements, a VPN implementation may consist of all software, all hardware, or a mixed solution.

Software-Based VPN Solutions

Software-based VPN solutions are used in deployments where high throughput is not required and budget is a concern. A software-based VPN solution consists of the following components.

VPN Client Software

VPN client software is normally installed on client desktop computers, laptop computers, and PDAs that require a secure LAN connectivity. Many operating systems, for example Microsoft Windows XP and Windows 2000, come with VPN software preinstalled and only require proper configuration. VPN client software enables all TCP/IP data transmission between the client computer and the VPN gateway to occur in encrypted form and provides authentication of the remote LAN user.

VPN Gateway Software

Similar to the VPN client software, most server operating systems, for example Microsoft Windows XP and Windows 2000 servers, come with VPN gateway software preinstalled and only require proper configuration. VPN gateway authenticates the remote VPN client and provides data privacy by transmitting all data in encrypted form.

Hardware-Based VPN Solutions

Hardware-based VPN solutions are mostly used for connecting two LANs over an insecure medium. These hardware devices are normally configured to authenticate each other, and usually no human-user authentication is performed to authenticate this connection.

Mixed VPN Solutions

The mixed VPN solution is the most prevalent form of VPN deployment. Mixed deployments use a software VPN client that is installed on user computers and a hardware-based VPN gateway installed at the remote LAN. Building VPN solutions in this manner provides high bandwidth at the gateway level and lowers the cost by using VPN client software.

Basic VPN Operation

The basic operation of VPN can be summarized as follows:

- VPN client and gateway are properly installed and configured to use the same encryption and authentication algorithms.
- A user account is created and allowed VPN connectivity. The user is provided with proper authentication information, for example the username and password, and the gateway IP address information.
- The user connects with the VPN gateway using the VPN client by providing username and password.
- The VPN gateway assigns an IP address to the VPN client, provides necessary TCP/IP parameters, and sets up the encryption parameters.
- The VPN client assumes the IP address assigned to it by the VPN gateway.
- When the client sends some data to the protected LAN, the VPN gateway performs the NAT function on the data and sends the data to the intended computer. Likewise, when a computer in the protected LAN sends data intended for the VPN client, the VPN gateway performs a reverse operation and sends the data to the VPN client.

Now that we are familiar with the two advanced security technologies, the 802.1X and the VPN technologies, let's use them to build a secure wireless LAN.

Building a Secure Wireless LAN with 802.1X and VPN Technology

In this example, we build a wireless LAN that consists of a wireless LAN user and an AP and that communicates with a wired LAN using a software-based VPN solution. The following are the network components that are necessary to build this LAN:

- A laptop computer equipped with ORiNOCO 802.11 Silver PC Card and Windows XP.
- A wireless LAN AP that supports the 802.1X protocol.
- A small Ethernet-wired LAN with a Windows 2000 Server and a desktop computer.

We further assume that the wired LAN is directly connected with the AP using one of the Ethernet jacks present in the rear of the AP, and that Alice and Bob are the users of the laptop and the desktop computers, respectively. Figure 10.1 shows us the desired configuration for the LAN equipment.

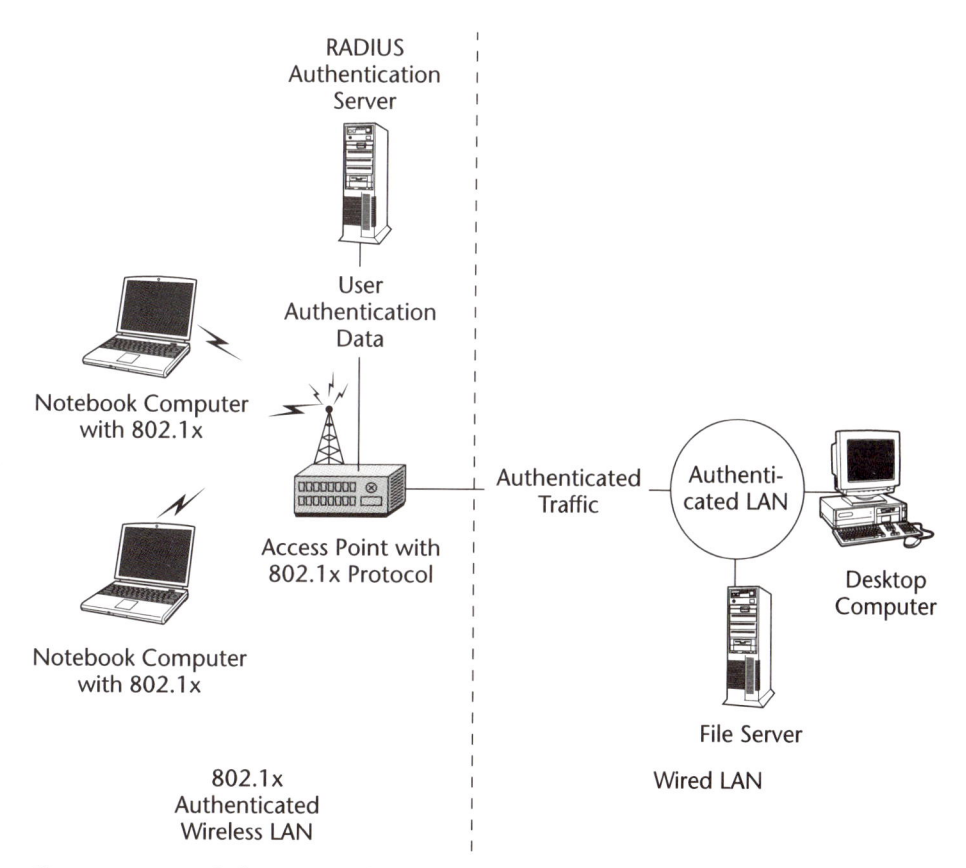

Figure 10.1 A wireless LAN with 802.11 authentication support.

Let's walk through the steps to build our secure wireless LAN that uses the robust 802.1X and VPN connectivity. We will first set up the LAN to use the 802.1X, and then we will add the VPN support to the LAN.

Setting Up the 802.1X for Wireless LAN

The 802.1X solution we are presenting here consists of a wireless LAN adapter with 802.1X software driver, an 802.11 AP with 802.1X support, and a wired LAN that is directly connected to the AP and consists of a RADIUS server and a desktop computer. In this example, we use Microsoft Windows 2000's Internet Authentication Service as our RADIUS server, Cisco 350 Series AP as the AP, and a client laptop computer equipped with ORiNOCO 802.11 Silver PC Card.

Configuring the RADIUS Server for the Wireless Users

Configuring the Windows 2000 Server's RADIUS service for use with our example server requires the following steps to be performed:

1. Click Start, point to Administrative Tools, and then point to Internet Authentication Service. Figure 10.2 shows the Internet Authentication Service screen.

- Right-click on Clients, and Select New Client.
- Enter a name for your access point and click on Next.
- Enter the IP address of your access point, and set a shared secret. Select Finish.
- Right-click on Remote Access Policies, and Select New Remote Access Policy.
- Name the policy EAP-MD5, and click on Next.
- Click Add. In this screen you're basically setting conditions for using EAP-MD5 to access the network (consult Windows 2000 documentation for more information on the exact restrictions that you can impose).
- Click on Edit Profile and select the Authentication tab. Figure 10.3 shows the authentication tab. Make sure Extensible Authentication Protocol (EAP) is selected. Deselect other authentication methods listed. Click OK.
- Windows asks you if you wish to view the Help topic for EAP; select No if you just want to get on with the installation. Click Finish.

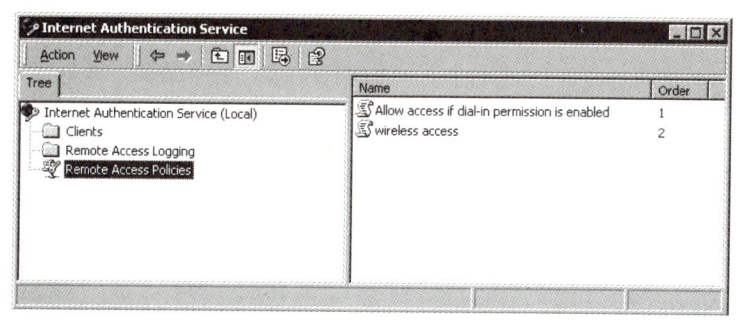

Figure 10.2 The Internet Authentication Service in Windows 2000.

Enabling Remote Access Login for Wireless LAN Users

- Click Start, point to Administrative Tools, and select Active Directory Users and Computers.
- Double-click on the user for which you want to enable authentication to bring up its account properties.

Select the Dial-in tab, and select Allow Access. Click OK.

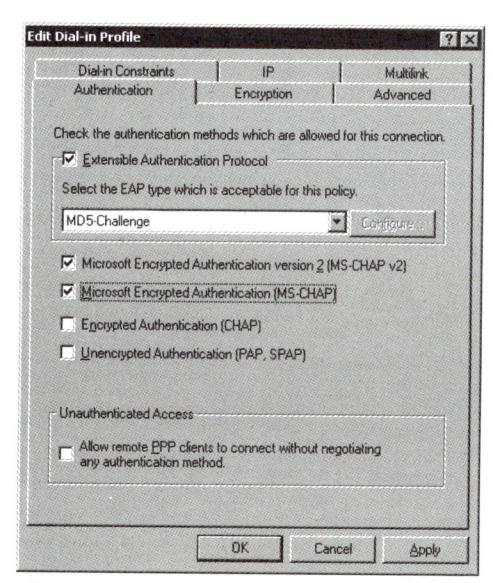

Figure 10.3 Windows 2000 Internet Authentication Service Authentication tab.

Configuring the Wireless LAN AP for 802.1X Authentication Protocol

You must configure the AP to use the RADIUS server. We assume that you have already performed the AP configuration using the Bob's desktop computer, which is connected to the AP via the wired Ethernet LAN.

We assume that you have set the proper SSID and channel on which the access point will operate and that you have taken the proper steps to secure the access point itself. These instructions use the Web management interface, although the identical configuration options are available from the terminal connection. It's important that you're running at least 11.08T firmware; as of this writing, the latest 11.10T is best. The following are the steps necessary to ensure proper setup of 802.1X:

CONFIGURING THE RADIUS SETUP

1. Log in to the AP Configuration setup using a Web browser.
2. From the home start screen, select Setup.
3. Select Security from under Services.
4. Select Authentication Server.
5. Under Server Name/IP, enter the IP address of the authentication server you've already set up with the Internet Authentication Service.
6. Server Type should be RADIUS, port 1812, and enter the shared secret that you set in step 5 of the server setup. Timeout can probably remain at the default 20 seconds, and ensure EAP Authentication is selected. Figure 10.4 shows the configuration screen of Cisco 350 Series AP 802.1X setup.
7. Select OK.

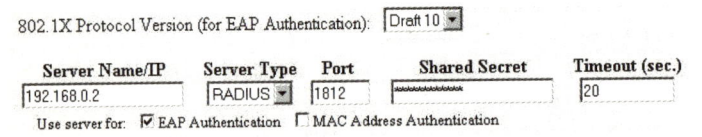

Figure 10.4 Cisco 350 Series AP 802.1X setup screen.

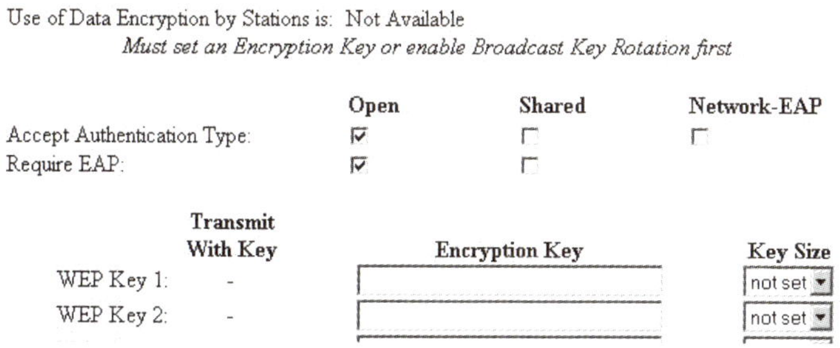

Figure 10.5 Cisco 350 Series AP WEP setup screen for EAP.

Enabling the 802.1X EAP Authentication

1. Go back to the Security screen. Select Radio Data Encryption (WEP). Figure 10.5 shows the WEP setup screen for Cisco 350 Series AP where you enable EAP.

2. Deselect all authentication types except for the Open options of Accept Authentication Type and Require EAP.

3. Select OK.

ENABLING ENCRYPTION

The only way to ensure strong mutual authentication between Windows XP and the access point is to enable dynamic WEP. Without it, your machines are vulnerable to a man-in-the-middle attack. 802.1X port access authentication isn't enough by itself.

1. Go back to the Radio Data Encryption (WEP) page.

2. Enter the encryption key, and select the appropriate key size.

3. Click OK.

4. Go to the Radio Data Encryption page once again.

5. Select Full Encryption from the Use of Data Encryption by Stations drop box as shown in Figure 10.6.

6. Click OK.

Figure 10.6 Cisco 350 Series AP WEP setup screen for encryption.

Configuring the Wireless LAN Adapter Software for 802.1X Protocol

For this task you should already be familiar with the steps required to install a wireless LAN adapter and the necessary software drivers; thus, we will examine only the configuration steps that are required for the 802.1X authentication support.

- Enabling 802.1X Authentication for Wireless Card:

1. Open up the properties for your wireless connection, either by right-clicking on My Network Places on the desktop and selecting Properties, or open up the Control Panel and select Network Connections (located under Network and Internet Connections if in Category View).

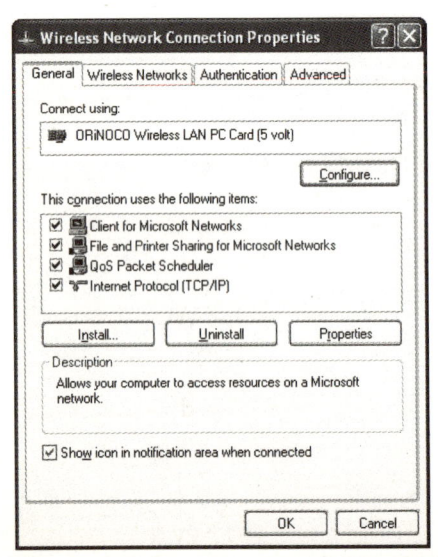

Figure 10.7 Wireless network connection properties under Windows XP.

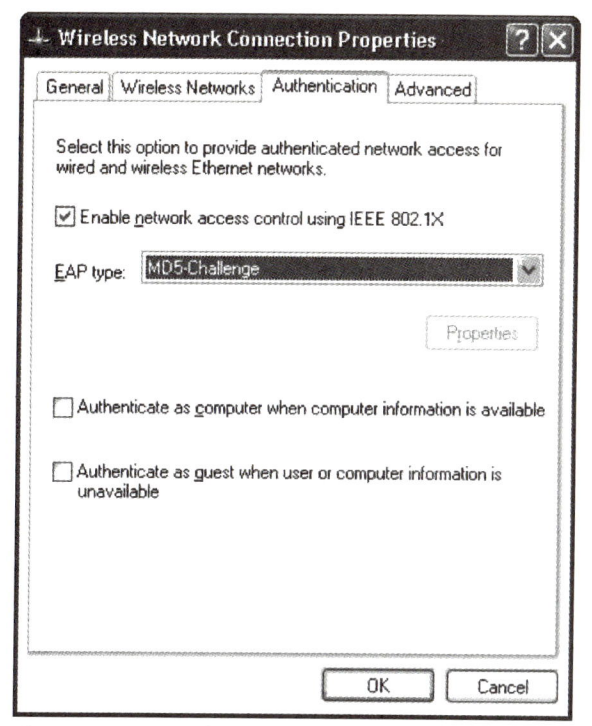

Figure 10.8 Wireless network authentication screen in Windows XP.

2. Right-click on the Wireless Network Connection, and select Properties. Figure 10.7 shows the wireless network connection properties.

3. Select the Authentication Tab, and ensure that Enable Network Access Control Using IEEE 802.1X is selected, and username/password-based EAP-MD5 is selected from the EAP type. Figure 10.8 shows the wireless network authentication screen in Windows XP.

ENABLING ENCRYPTION

1. To enable encryption for a wireless network, click on the Wireless Networks tab.

2. Select the wireless network on which you want to enable dynamic WEP from under Available Networks, and select Configure. Figure 10.9 shows the WEP configuration screen in Windows XP.

3. Select Data encryption (WEP-enabled), and ensure The Key is Provided for Me Automatically is also selected.

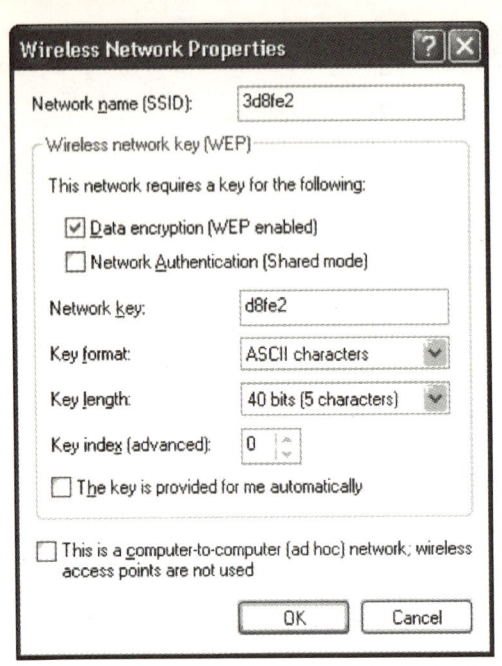

Figure 10.9 WEP encryption configuration in Windows XP.

Adding VPN Connectivity to Provide Higher Security

The preceding steps described how to improve WEP support, as defined in the basic 802.11 wireless LAN, by using the 802.1X authentication protocol. Adding VPN connectivity provides an additional layer of security that complements the security provided by the 802.1X protocol. In this section, we present an example of setting up VPN connectivity between a wireless LAN client computer installed with Microsoft Windows 2000 OS and a computer on the wired LAN installed with Microsoft Windows 2000 Server.

Setting Up Windows 2000 VPN Gateway/Server

Configuring Windows 2000 server for using as a VPN server includes the following steps:

1. Install and enable VPN. Most of the VPN server components are preinstalled on the Windows 2000 server; still, you need to install some components and enable the VPN server.

2. Configure the VPN Server. You also have to configure the security parameters for Point-to-Point Tunneling Protocol (PPTP), which provides

data encryption using Microsoft Point-to-Point Encryption and the Layer Two Tunneling Protocol (L2TP) that provides the data encryption, authentication, and integrity using IPSec protocol.

3. Set up users to access the VPN. You will have to set up the VPN server to allow the users you want to grant VPN access.

4. Let's get started with setting up a Windows 2000 server as a VPN server.

Installing and Enabling VPN

To install and enable a VPN server, follow these steps:

1. On the Microsoft Windows 2000 VPN Server, confirm that the connection to your local area network (LAN) is correctly configured.

    a. Click Start, point to Administrative Tools, and then click Routing and Remote Access.

    b. Click the server name in the tree, and then click Configure and Enable Routing and Remote Access on the Action menu. Figure 10.10 shows the Routing and Remote Access Screen in Windows 2000. Click Next.

2. In the Common Configurations dialog box, click Virtual private network (VPN server), and then click Next.

3. In the Remote Client Protocols dialog box, confirm that TCP/IP is included in the list, click Yes, All of the Available Protocols Are on This List, and then click Next. Figure 10.11 shows the Remote Client Protocols dialog box.

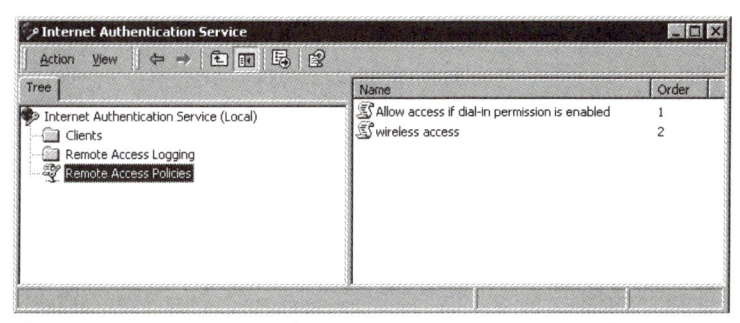

Figure 10.10 Routing and remote access screen in Windows 2000.

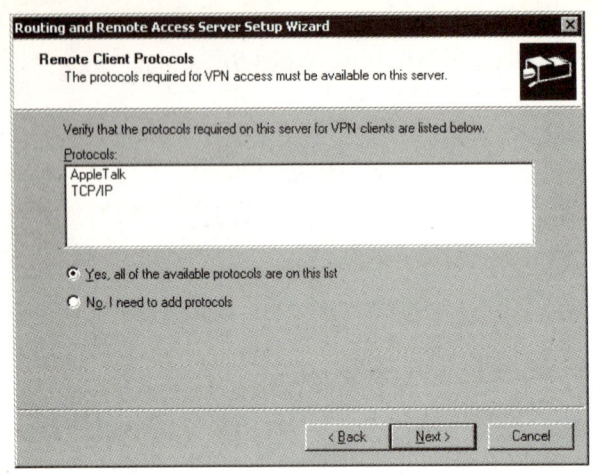

Figure 10.11 Remote Client Protocols dialog box showing the client protocols.

4. In the IP Address Assignment dialog box, select Automatically in order to use the DHCP server on your subnet to assign IP addresses to dial-up clients and to the server.

5. In the Managing Multiple Remote Access Servers dialog box, confirm that the No, I don't want to set up this server to use RADIUS now check box is selected. Click Next, and then click Finish.

6. Right-click the Ports node, and then click Properties.

7. In the Ports Properties dialog box, click the WAN Miniport (PPTP) device, and then click Configure.

8. Type the maximum number of simultaneous PPTP connections that you want to allow in the Maximum Ports text box. The maximum number may depend on the number of available IP addresses. For example, if you want to use only 25 IP addresses, enter 25 for the maximum number of simultaneous PPTP connections.

9. In the Ports Properties dialog box, click the WAN Miniport (L2TP) device, and then click Configure.

10. Type the maximum number of simultaneous L2TP connections that you want to allow in the Maximum Ports text box. The maximum number may depend on the number of available IP addresses. For example, if you want to use only 25 IP addresses, enter 25 for the maximum number of simultaneous PPTP connections.

Configuring the VPN Server

To configure the VPN server, follow the steps in the following paragraphs.

Configuring the Remote Access Server as a Router

For the remote access server to forward traffic properly inside your network, you must configure it as a router with either static routes or routing protocols so that all the locations in the virtual LAN are reachable from the remote access server. Follow the steps that follow to configure the server as a router.

Click Start, point to Administrative Tools, and then click Routing and Remote Access.

1. Right-click the server name, and then click Properties.
2. On the General tab, click to select Enable This Computer As A Router.
3. Select Local Area Network (LAN) Routing Only. Click OK to close the Properties dialog box.

Setting Up Addresses and Name Servers

The VPN server must have IP addresses available in order to assign them to the VPN server's virtual interface and to VPN clients during the IP Control Protocol (IPCP) negotiation phase of the connection process. The IP address assigned to the VPN client is assigned to the virtual interface of the VPN client.

For Windows 2000-based VPN servers, the IP addresses assigned to VPN clients are obtained through DHCP by default. You can also configure a static IP address pool. The VPN server must also be configured with name resolution servers, typically DNS and WINS server addresses, to assign to the VPN client during IPCP negotiation.

Setting Up Users for VPN Access

By default, users are denied access to dial-up. Configure the dial-in properties on user accounts and remote access policies to manage access for dial-up networking and VPN connections.

VPN Access by User Account

If you are managing remote access on a user basis, click Allow Access on the Dial-In tab of the user's Properties dialog box for those user accounts that are allowed to create VPN connections. Delete the default remote access policy called "Allow Access If Dial-In Permission Is Enabled." Then create a new remote access policy with a descriptive name, such as "VPN Access If Allowed By User Account." For more information, see Windows 2000 Help. If the VPN server is also allowing dial-up remote access services, do not delete the default policy, but move it so that it is the last policy to be evaluated.

242 Chapter 10

VPN Access by Group Membership

If you are managing remote access on a group basis, click Control Access through Remote Access Policy Radio on All User Accounts. Create a Windows 2000 group with members who are allowed to create VPN connections. Delete the default remote access policy called Allow Access If Dial-In Permission Is Enabled. Next, create a new remote access policy with a descriptive name such as "VPN Access If Member of VPN-Allowed" group, and then assign the Windows 2000 group to the policy. If the VPN server also allows dial-up networking remote access services, do not delete the default policy; instead move it so that it is the last policy to be evaluated.

Configuring the VPN Client

Follow these steps to set up a connection to a VPN:

1. Log in as the administrator on the client computer. This option is available only if you are logged on as a member of the Administrators group.

2. On the client computer, confirm that the connection to the wireless LAN is correctly configured.

3. Click Start, point to Settings, and then click Network And Dial-Up Connections.

4. Double-click Make New Connection. This will start the Network Connection Wizard. Click Next. The Network Connection Type screen will appear as shown in Figure 10.12.

5. Click on Connect To A Private Network through The Internet, and then click Next.

6. Click Do Not Dial The Initial Connection. Click Next.

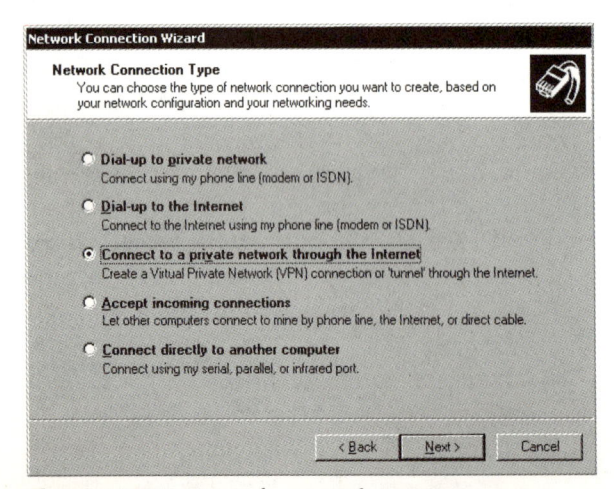

Figure 10.12 Network connections type screen.

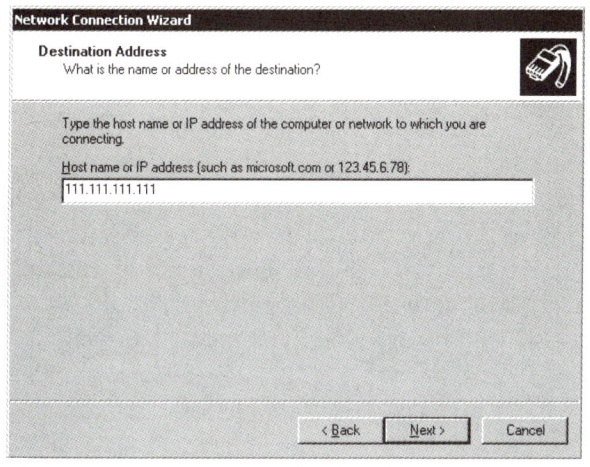

Figure 10.13 VPN server identification settings screen.

7. Type the host name (for example, vpn.acme.com) or the IP address (for example, 111.111.111.111) of the computer to which you want to connect, and then click Next. Figure 10.13 shows the VPN server identification settings screen.

8. Click to select For All Users if you want the connection to be available to anyone who logs on to the computer, or click to select Only for Myself to make it available only when you log onto the computer. Click Next.

9. In Completing the Network Connection Wizard screen, type a descriptive name for the connection, and then click Finish. Completing the Network Connection Wizard screen is shown in Figure 10.14.

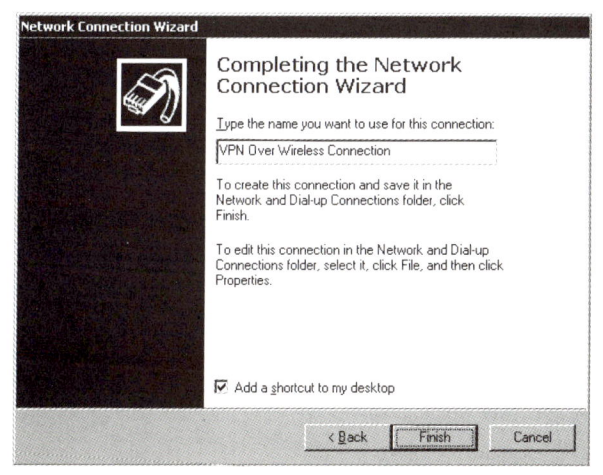

Figure 10.14 Completing the Network Connection Wizard screen.

Testing the VPN Connectivity

To test the VPN connectivity, follow these steps:

1. Click Start, point to Settings, and then click Network And Dial-Up Connections.
2. Double-click the new connection you just created.
3. The VPN server should prompt you for your username and password. Figure 10.15 shows the VPN connection window. Enter your username and password, click Connect, and your network resources should be available to you in the same way they are when you connect directly to the network.

Point-to-Point Wireless Connectivity between Two Sites

In campus-type business settings where an office may be split into more than one building, the individual LANs in each building can be connected with each other using the 802.11 standard wireless LAN technologies. In this section, we talk about using the OriNOCO 802.11 compliant, point-to-point wireless kit to provide point-to-point wireless connectivity between two physically separated sites. We first talk about the basic requirements for setting up wireless connectivity between two sites, then we discuss the basic network configuration of this connectivity, and finally we walk you through the basic steps involved in setting up such connectivity to give you a general idea of the effort involved.

Figure 10.15 VPN connection window.

Point-to-Point Wireless Connectivity Requirements

Point-to-point wireless connectivity normally requires the following:

Obstruction-free space. The point-to-point connectivity requires that there be no obstructions, for example other buildings or trees, between the sites that are connected using the point-to-point wireless LAN technology. ORiNOCO claims that their point-to-point kit works within a range of six miles without any obstruction between them.

High security. If a point-to-point wireless connection is not secured, the data can be picked up by the adversary. We strongly suggest that you use the best available security option when establishing point-to-point connectivity.

Network Configuration

As mentioned earlier, the point-to-point wireless connectivity that we are discussing in this example includes two separate wired LANs that need to be connected with each other, VPN gateways to provide network level security, and the ORiNOCO wireless point-to-point kit. The entire configuration is shown in Figure 10.16.

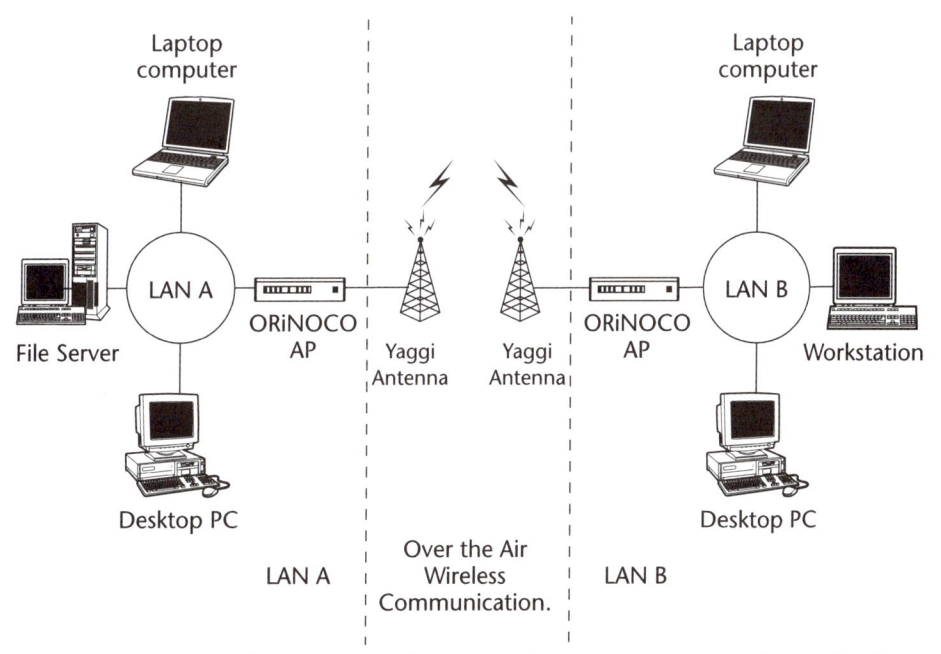

Figure 10.16 Network configuration for connecting two LANs using ORiNOCO hardware.

Setting Up ORiNOCO Point-to-Point Radio Backbone Kit

The Point-to-Point Radio Backbone Kit is an easy-to-install, highly reliable 11-Mbps building-to-building wireless LAN connectivity solution. The kit includes all the necessary hardware, software, and management components needed to establish a license-free 2.4-GHz wireless LAN bridge that spans up to 6 miles under clear line–of–sight conditions. The kit contains the following important components.

- Two OR-500 Outdoor Routers.
- Two ORiNOCO gold cards.
- Two standard Ethernet cables. For connecting the routers with the local LANs at each site.
- Two pigtails.
- Two 50-foot lengths of LMR low-signal-loss cable (to connect the indoor installation to the outdoor antenna).
- Two surge arresters. The ORiNOCO Surge Arrester is an indispensable part of your outdoor antenna installation, to protect your sensitive electronic equipment from transients or electrostatic discharges at the antenna.
- Two-14.5 db Yagi antennas. The actual antennas that transmit the radio waves in the air.

In this section, we talk about the steps that will help you understand the basic process. For information on actual steps, please visit Agere Corporation at http://www.Orinoco.com.

Antenna Installation

The foremost step in establishing wireless connectivity is to ensure that wireless LAN connectivity between proposed sites is, in fact, feasible and realistic. Following are some of the basic site-related issues that you should consider when performing site survey and locating the best places for antenna installation:

Distance between the two sites. ORiNOCO kit claims that it will function up to a range of six miles between the sites without any obstruction. If the distance is more than six miles, you might want to contact Agere or other vendors to explore other possible options.

Nature of obstructions between the two sites. If buildings or other concrete objects are present between the two sites you want to connect using the wireless connectivity, it might not work. Trees between the two buildings might not cause the wireless connectivity to fail, but may result in a degraded LAN performance.

Line of sight. When locating a suitable place for antenna installation, make sure that the two antennas are installed in the line of sight. This will ensure that the radio signals can reach from one antenna to the other.

Distance between routers and antennas. The ORiNOCO kit does require use of router device at each site (OR-500 Outdoor Routers) that is installed indoors. The length of cable between the antenna and the router is a decisive factor when considering LAN performance. Make sure that you install the antennas such that the distance between the outdoor router and the antenna is minimal.

Consult the installation guide for more information on the actual installation process.

Installing the Outdoor Routers

The hardware of ORiNOCO Outdoor Router device is designed for indoor mounting and operation. The ideal location to install the outdoor router unit must satisfy the following requirements:

- The location must allow for easily disconnecting the Outdoor Router unit from the AC wall outlet.
- The location should provide a connection to the network backbone that may either be the Ethernet LAN cable that connects it to a hub, bridge, or directly into a patch panel or the wireless connection via a second ORiNOCO PC Card that is inserted into the other PC Card slot of the Outdoor Router device.
- The location should be as close as possible to the point where the antenna cable will enter the building.

The following provides basic steps that you must follow to install the ORiNOCO outdoor routers at each site. For complete information, please consult the ORiNOCO Point-to-Point Backbone Kit.

1. Install the outdoor routers at each location according to the requirements mentioned above.
2. Insert the ORiNOCO Gold Card PC Card in the router device PC Card slot.
3. Connect the cable from each antenna to the router at each location.
4. Connect the router device with local LANs.
5. If all devices have been properly installed according to the instructions in the ORiNOCO Point-to-Point Backbone Kit installation manual, you are done with the physical installation steps.

6. Power up the routers at each location. The next step will involve actual configuration of the router devices and fine-tuning the antenna to obtain best performance.

7. Use the Router Client License Kit to configure the router per instructions provided in the license kit. Also, do not forget to turn on the encryption parameters; we suggest that you do not use this wireless LAN connectivity solution without the encryption option.

8. Use the wireless signal monitoring software that comes with the kit to ensure that you are receiving signals from LANs at each site. If you do not receive the signal from the other site, check the installation steps.

9. If both sites seem to be installed and configured properly, using a computer on one of the LANs, use the TCP/IP-ping application program to test and ensure that computers in the LANs connected using the kit can communicate with each other. For example, if the two sites you wirelessly connected are called site A and site B, respectively, and you know that one of the computers on site B has a fixed IP address of 192.168.0.16, try using a computer on site A that is connected to site B over the wireless link and ping the computer with IP address 192.168.0.16. If you are using Microsoft Windows-based computer, using the ping application program may look like following:

```
C:\>ping 192.168.0.16
Pinging 192.168.0.16 with 32 bytes of data:
Reply from 192.168.0.16: bytes=32 time<10ms TTL=128
Reply from 192.168.0.16: bytes=32 time<10ms TTL=128
Reply from 192.168.0.16: bytes=32 time<10ms TTL=128
Reply from 192.168.0.16: bytes=32 time<10ms TTL=128
```

If there is an error or the computer is not configured correctly, you might get an error as follows:

```
C:\>ping 192.168.0.16
Pinging 192.168.0.16 with 32 bytes of data:
Request timed out.
Request timed out.
Request timed out.
Request timed out.
```

10. If you get the error message, check the operating system instructions and the instructions provided by the network card provider. If you do not get any error messages, it ensures that your wireless LAN connectivity is operational.

Securing the Point-to-Point Wireless Connectivity Using VPN

Securing the point-to-point wireless connectivity using VPN requires the same steps as mentioned in the section *Adding VPN Connectivity to Provide Higher Security*. We strongly recommend that you set up VPN connectivity between two points to ensure a high level of security.

Secure Remote Access from a Wireless LAN over the Internet Using VPNs

Remote access over the Internet from a wireless LAN usually includes a scenario in which a remote user accesses a private LAN, for example a LAN at a corporate office, using his or her computer in a wireless LAN environment. The wireless LAN environment here could be a home-based wireless LAN or a public wireless LAN hotspot. The security needs of such remote access are extremely high because both the remote LAN and the computer accessing the remote LAN can be subject to an attack. We strongly recommend that a LAN must never allow a remote access without VPN connectivity.

A basic network that provides secure remote access over the Internet consists of the following components as shown in Figure 10.17.

- Remote LAN installed with VPN gateway connected with the Internet.

- Wireless LAN consisting of one or more APs and computers with wireless LAN adapters, with at least WEP security-enabled.

- Wireless LAN connected with the Internet either using a broadband connection (DSL or digital cable) or dial-up modem.

- VPN client software installed on the client computers accessing the remote LAN.

The steps involved in establishing the secure connectivity in this scenario include the steps that are required to set up the wireless LAN gateway and client computers as described in section *Adding VPN Connectivity to Provide Higher Security* earlier in this chapter.

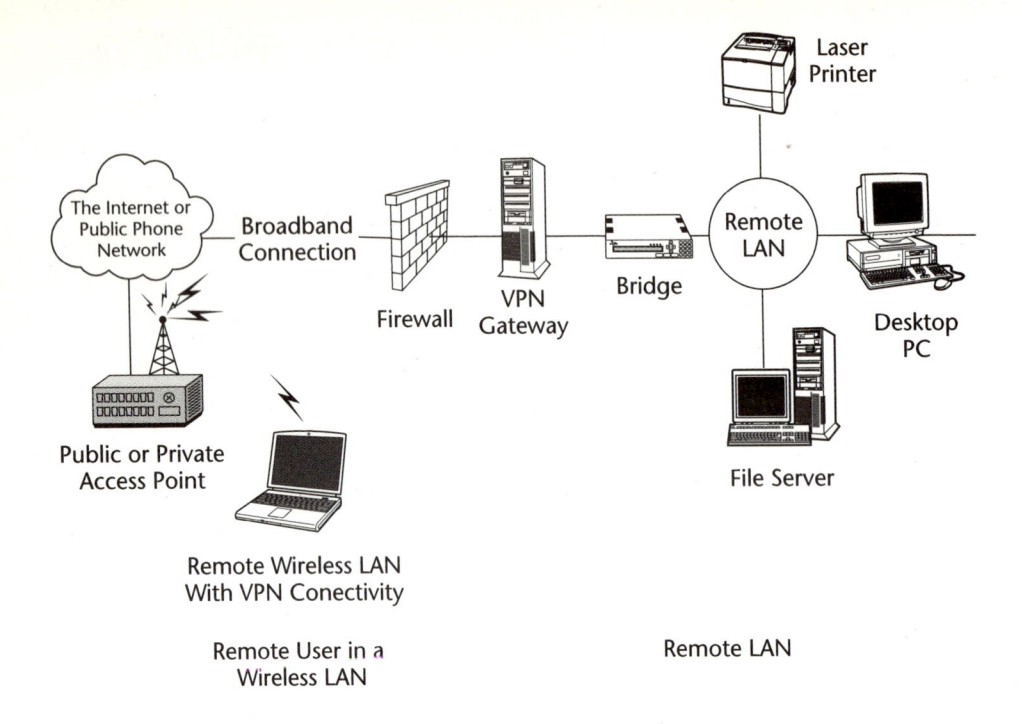

Figure 10.17 Secure wireless access to a remote site over the Internet.

Summary

In this chapter, we used the wireless LAN installation expertise that we built in the last chapter. We further added advanced security features to the base of our wireless LANs that we built in the last chapter. We discussed the 802.1X authentication protocol and VPNs and described how they can be added to an existing wireless LAN to provide security and answer the problem that WEP could not provide. We also talked about how two physically separated sites can be interconnected using wireless LAN technology and how remote wireless LAN users can connect to a remote site.

In the next chapter, we talk about the basic problems you might encounter when deploying a wireless LAN and simple solutions that can alleviate them. We also talk about optimizing wireless LANs through monitoring wireless LAN performance.

PART Four

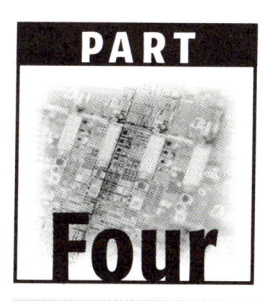

Troubleshooting and Keeping Your Wireless LAN Secure

Though building a wireless LAN is a challenging task, troubleshooting and ensuring its security can be a nightmare. This part of the book provides you with a general guideline that can help you identify potential problems with your wireless LAN and keep the wireless LAN secure.

Chapter 11 discusses the most commonly known problems and their solutions. Generally speaking, once you identify these problems, the solutions are relatively easy to implement.

In Chapter 12, we discuss the process that may be followed when establishing a security policy. We talk about assessing your security policy needs and creating a policy that fulfils those needs. We also provide you with a sample policy that you can use as a guide.

CHAPTER 11

Troubleshooting Wireless LANs

Although wireless LANs are easy to install and set up, troubleshooting them can be a nightmare. You will likely encounter troubles with an 802.11 deployment because of unplanned environmental changes, for example office relocation or change in the physical setting, utilization growth, security breaches, and improper configuration parameters.

In this chapter, we talk about some of the issues surrounding the troubleshooting maintenance of a wireless LAN. These issues include common problems, such as handling bandwidth congestion due to competing devices, upgrading wireless LAN equipment, and optimizing and managing network overload through monitoring.

Common Problems

Typical wireless LAN hardware consists of PC Card wireless LAN adapters and wireless LAN access points (APs). The APs are quite resilient when it comes to hardware problems as they reside at one location and are normally not handled by users. However, PC Cards are vulnerable to a number of hardware and software problems because they are removable devices and users of computer notebooks and laptops often remove them to use other PC Card devices. In this section, we examine the most common hardware- and software-related problems that wireless LAN adapters and APs suffer from.

Hardware Problems

Hardware problems can be easily recognized through better understanding of the wireless LAN equipment. This includes becoming familiar with the model and features of the products, the functions of light indicators on the AP and wireless LAN adapters, specific technologies involved, and their installation procedures. The paragraphs that follow describe some of the common hardware-related problems.

Disconnected AP Power or Network Cable

Access points normally suffer from this problem where the electric power to the APs gets disconnected due to a human mistake. This problem is easy to detect, as APs without power will not send any periodic beacon, which they normally would send out.

Inoperable Technology Use

If an 802.11b wireless LAN card is used in an 802.11a wireless LAN, it will not work. The problem can be diagnosed by carefully examining the physical wireless LAN adapter to ensure that it is compatible with the AP in the LAN.

Antenna Diversity

If the antenna of an AP gets broken, the AP might not work at all or may provide extremely poor performance. APs should be monitored periodically to ensure that the antennas are properly installed.

Older Firmware

Often wireless LAN manufacturers publish newer firmware, which can be uploaded into the wireless LAN adapter or the AP. Sometimes only one of the components, either the wireless LAN adapter or the AP, is upgraded with new firmware and the other device is not, making both incompatible and inoperable with each other. Any time a firmware upgrade is performed, all other devices should be made checked to ensure compatibility with the new firmware.

Equipment Failure

In case of an equipment failure, you may need to quickly provision and configure new hardware components. To ensure that this can be done in a timely manner, you must record all configuration and customization information, for example AP passwords and WEP keys, about your APs and other wireless LAN equipment.

Mishandling of Wireless LAN Adapters

Most mobile computers use PC Card-based wireless LAN adapters. PC Cards are sensitive hardware devices. Often they get dropped on the floor by mistake, resulting in internal circuit breakdown. Often signs for such breakdown occur when a computer might not recognize the card at all or gets hung when booted. If such a problem is noticed, try using a different wireless LAN card on the computer where the problem is noticed. If the other card does not demonstrate this problem, it's likely that the first card has gone bad.

Cable Problems

Wireless APs are radio devices, and all cables connecting to them must be of high quality. Use high-quality shielded cables to ensure that there are no unintended radio frequency leakages to interfere with the AP radio signals. The antenna cables should not be longer than specified by the wireless LAN device manufacturer. The longer the antenna cable, the lower the signal strength; consequently, the lower the range. To ensure cable-related interference problems, you should avoid the following:

- The connections to either end of the antenna cable must be secured. Loose cables result in degraded radio quality.
- Avoid using cables that show physical signs of damage. Antenna cables with obvious physical damage can result in degraded radio signals.
- Avoid shared antennas and power cable runs. Power cables can produce electromagnetic interference that may affect the signal in the antenna cables within close proximity.

Effect of Building Material on Wireless LANs

The building material at a wireless LAN site affects the performance of the LAN. The following table illustrates the maximum number of walls that can be present between access points and wireless LAN adapters without any degradation of the coverage.

Table 11.1 The Effect of Building Material on Wireless LAN Performance.

MATERIAL TYPE	MAXIMUM WALLS BETWEEN AP AND AN
Paper or vinyl walls	No impact on signal
Solid or precast concrete walls	1 or 2
Concrete block	3 or 4
Wood or dry wall	5 or 6
Metal walls	0

Software Problems

Because APs are always fewer in a wireless LAN compared to the number of wireless LAN adapters in a wireless LAN, the software problems are more common among wireless LANs. Some of these problems may be discovered when the wireless LAN adapter is set up, while others may arise any time during the use of a wireless LAN. Some of the common software-related problems are described in the paragraphs that follow.

Incorrect Configuration

Wireless LAN adapters do not have many configuration problems, so it is easy to overlook them when configuring a wireless LAN adapter. The common configuration problems include incorrect selection of wireless LAN operation mode and channel settings. For example, if the LAN is operating in infrastructure mode and the adapter is configured for the ad-hoc mode, the LAN adapter will not be able to participate in the wireless LAN.

WEP Encryption Setting

If a wireless LAN AP is configured to use WEP encryption and the wireless LAN adapters are not, the network will not function because the wireless LAN AP will be communicating in the encrypted mode, whereas wireless LAN adapters will be communicating in unencrypted mode. To make sure that such a problem does not arise, you should make sure that anyone responsible for or capable of modifying the adapter settings is aware of the encryption mode used in the wireless LAN.

Incorrect Extended Service Set Identifier (ESSID)

In infrastructure mode, each AP is assigned an ESSID. All wireless LAN adapters that need to communicate with an AP, or group of APs, must be configured with the correct ESSID. You need to further ensure that if multiple wireless LANs are within the radio range and you want to isolate them, you must use a unique ESSID.

Software Installation Problems

If an incorrect version of software or the wrong software driver for wireless LAN adapter is installed on a computer, the adapter will not function at all and the computer will not be able to participate in the wireless LAN. To ensure that you are installing the correct software drivers, consult the wireless LAN adapter manuals or contact the manufacturer.

Authentication Problems

If you are using the 802.1X authentication protocol, you should make sure that you are using the correct credentials to prove your identity. If your wireless LAN uses username and password-based authentication, you should make sure that you type them correctly.

VPN-Related Problems

Using VPNs can become quite cumbersome if the instructions for VPN setup are not properly provided to the VPN users. You should make sure that all users fully understand the purpose and procedure of using VPNs, as VPNs require configuration of various cryptographic protocols and algorithms, which might not be the most trivial things to understand.

Network Design and IP Address Assignment

If an access point or a client adapter becomes inaccessible on the network, you need to ensure that another computer in the LAN is not sharing the IP address. If you have multiple LANs, you need to further ensure that the network is properly designed and correct gateway information is configured in all appropriate places, for example routers and network adapters. (See Chapter 1 for more information on network gateways.)

Handling Bandwidth Congestion Due to Competing Devices

All radio frequency-based devices suffer interference problems due to other devices operating in the same frequency. When different types of devices operate in the same physical area, they may not cooperate with each other in terms of bandwidth usage. Most wireless communication protocols require that, when they send a data packet to another wireless device, the receiving device must send an acknowledgment packet confirming the receipt of the data packet; otherwise it resends the data packet. For example, if a wireless LAN adapter sends a data packet to an AP, the AP must reply upon successful receipt of the data packet; if the AP does not reply with an acknowledgment packet, the wireless LAN adapter keeps on retrying by retransmitting the packet. The retransmission works under most circumstances, but if there is too much interference and data is not able to travel among the wireless LAN devices, they all become busy sending out packets and retrying, which quickly makes the wireless LAN bandwidth congested with retransmissions. Devices that operate in the 2.4-GHz Industrial, Scientific, and Medical (ISM) unlicensed band, for example, 802.11b, are especially vulnerable to this problem. Following are some of the devices that operate in the 2.4-GHz band; their communication may disrupt the operation of an 802.11b-based wireless LAN.

- **Bluetooth.** Bluetooth-based wireless devices are short-distance radio devices that use 2.4-GHz band. Bluetooth devices are PDAs and small wearable electronic gadgets and require very little bandwidth, but their fast growth and RF band competition is adding a potential threat to the 802.11b wireless LANs.
- **Microwave ovens.** Many microwave ovens also operate in 2.4-GHz band and cause interference and bandwidth congestion in a wireless LAN.
- **Cordless phones.** The 2.4-GHz cordless phones are also known to disrupt the 802.11b wireless LAN operation.

The following are some of the best ways to avoid bandwidth congestion due to competing devices.

- Remove and minimize all devices that operate in the RF band of the wireless LAN.
- Users of the LAN should be informed that the LAN performance would be greatly degraded if they use devices that use the same bandwidth as the wireless LAN.
- Another option might be to use the 802.11a-based wireless LAN equipment. 802.11a equipment operates in the newly licensed 5-GHz band.

The availability of fewer devices in the 5-GHz band makes it less vulnerable to the RF interference and bandwidth congestion.

- Additionally, you can experiment with relocating the AP and all competing devices. Switch to channel 1 or channel 11 on the access points. For example, if the computer with a wireless LAN adapter is near a cordless-phone base station, try using a card that allows you to use a remote external antenna.

Upgrading Wireless LANs

Upgrading wireless LANs is a challenging task, as it requires a careful assessment of your wireless needs and the benefits of upgrading to the new technology. You should consider the following before planning to upgrade to a new wireless LAN technology:

The RF band of the new wireless LAN that you are willing to deploy. If the RF band of the new wireless LAN is already too congested with competing devices, for example 900 MHz, you should think again. It is never a good idea to upgrade to a technology that utilizes an RF band that is already quite busy.

The claimed wireless LAN speed by the manufacturer. If you are upgrading your wireless LAN because of the speed claims by new wireless LAN manufacturers, you should ensure that the claims are true and they match your requirements. For example, wireless LAN equipment based on 802.11a has a smaller range than its rival 802.11b. If operating range was more important than the LAN speed, you should rethink your upgrade plans.

The security features available in the wireless LAN technology. You should make sure that the new technology provides sufficient security features and meets your security requirements.

Upgrading. Do all users need to be upgraded to the new technology? Under most circumstances, not all users and LAN devices need high-speed connections. Understanding the speed need will help you decide if it is a good idea to upgrade the wireless LAN.

Migrating to the new technology. Are your upgrade needs immediate or can you wait and slowly migrate to the new technology? A few manufacturers, for example Agere Systems, make wireless LAN APs that can provide both 802.11b and 802.11a wireless AP services. If you already have a wireless LAN that uses 802.11b and you want to upgrade to 802.11a wireless LANs, you might want to consider deploying such wireless LAN APs and slowly provide users with new 802.11a devices.

Optimizing and Managing the Network Load through Monitoring Wireless LAN Quality

Your wireless LAN may exhibit various different types of loads within the same physical site. There might be some areas where the wireless LAN would be heavily used and provides relatively weak performance due to the fact that APs are overloaded. A regular monitoring of wireless LAN quality can help you decide whether you need more APs at a given site or you can simply change the physical installation locations of some of the APs to balance the LAN quality. Monitoring the LAN quality on a regular basis ensures that you are accounting for any possible changes in the LAN needs of your users.

Summary

Wireless LANs are relatively trouble-free when compared with wired LANs. The most common problems in a wireless LAN arise due to incorrect configuration and setup of the wireless LAN equipment. It is a good idea to gain as much familiarity with the wireless LAN technology and products as possible. The users of wireless LANs should be made aware of the common parameters of wireless LANs. For example ESSID, encryption parameters, and the operation mode should be clearly communicated.

The RF band used by wireless LANs should be carefully guarded. LAN users should be discouraged from using devices in the wireless LAN coverage area that can potentially disrupt or degrade a wireless LAN performance. When upgrading a wireless LAN, you should consider the radio frequency, speed, and security of the available options. You should prefer a migration strategy to a replacement when upgrading a wireless LAN technology. Wireless LANs should be regularly monitored for the quality of the service they provide. Under most circumstances, adding new APs, and relocating existing APs within a geographical region can improve LAN performance.

In the next chapter, we talk about keeping your wireless LAN secure by establishing security policies, making sure that security policies are complied with, and making sure that proper measures are taken to ensure intrusion detection and containment.

CHAPTER 12

Keeping Your Wireless LAN Secure

Despite the constant increase in security features of wireless LAN products and technology, the risk of attack and penetration remains high. As with wired networks, it is only a matter of time before someone breaches the security on your wireless network. Understanding the criminals' goals, tricks, and techniques will help ensure that you and your wireless devices and network remain secure and one step ahead of them. Wireless LANs must be secured against attacks from both hackers and improper use. Besides ensuring that you take the best measures against any possible attack on the network, wireless security experts agree that a strict security policy may help reduce the vulnerability of wireless LANs.

It is a good idea to understand how to develop and integrate an effective wireless security policy into your enterprise to ensure wireless LAN continuity. In this chapter, we talk about developing practical wireless LAN security policies that work. We discuss the process of developing and establishing wireless LAN security policies and how to integrate them into an organization.

Establishing Security Policy

A wireless LAN security policy establishes information security requirements for a deployment to ensure that confidential information and technologies are not compromised and that network resources and other computing devices are protected.

In order to establish a successful security policy, you must understand your security policy requirements, create the policies, and deploy them carefully by announcing them among the LAN users.

Understanding Your Security Policy Requirements

Your security policy requirements are often dictated by the threats that you need to secure your wireless LAN against. Threats that a wireless LAN deployment may be vulnerable to depends, at least, on the deployment scenario (for example large enterprise and government wireless LANs might be of higher interest to an adversary); the confidentiality of the data in the wireless LAN (for example, a LAN containing financial data would be more vulnerable than a LAN containing publicly available information on Shakespeare's *Romeo and Juliet*); the physical location (for example, a wireless LAN located in the middle of nowhere would be difficult to reach compared to a wireless LAN in the middle of a city); and the LAN resources (for example, a high-bandwidth Internet connection would be more appealing to a hacker than a LAN that is not connected to the Internet).

When creating a wireless LAN security policy, you should consider, at least, user authentication, data privacy, measures against known wireless LAN attacks, AP configuration parameters, client-side configuration risks, and measures against war driving as the primary requirements of your wireless LAN security.

Authentication

Uncontrolled wireless access can allow attackers to read email, sniff passwords, gain administrative access to machines, plant access to machines, plant Trojan horses or back doors, and use wireless access points to launch other attacks. A wireless LAN security policy must require an adequate level of authentication to ensure that most possible threats are minimized.

Data Privacy

The data in a wireless LAN is vulnerable to tampering and spoofing. An adversary within the range of wireless LAN radio waves can monitor the LAN traffic and intercept the data. If the data is not encrypted, the adversary can easily modify the data or gain access to confidential information. A good security policy will require that all data transmission over a wireless LAN must only take place in encrypted form. Also, any confidential data must never be exchanged over a wireless LAN.

Measures Against Attacks on Wireless LAN

A wireless LAN security policy must include provisions to deter attacks on the wireless LAN. It must address, at least, the following known attacks. See Chapter 6 for more possible attacks on wireless LANs.

Wireless Device Insertion Attacks

The insertion attack on a wireless LAN is conducted by a hacker or an adversary by placing or brining a wireless LAN device well within the range of a wireless LAN. If a wireless LAN is not properly configured, the adversary can make the wireless LAN believe that the LAN device he or she introduces is a legitimate client of the wireless LAN and gain access to the LAN. There are two common attacks on wireless LANs:

Unauthorized Wireless LAN Clients. Unauthorized wireless LAN clients are mobile computers or other computing devices that have a wireless LAN adapter installed and can forge a LAN user to gain access to the LAN.

Enforcing MAC-level and the use of 802.1X-based authentication can deter the insertion attacks by unauthorized wireless LAN clients.

Rogue APs. Hackers may also place a wireless LAN AP within the operating range of a wireless LAN to impersonate a real AP. In this case, the wireless LAN adapters may be fooled into believing that the rogue AP is, in fact, a legitimate AP. The rogue AP operator, the hacker who installs a rogue AP, can easily gain authentication information from users when they authenticate themselves to the AP. Once the hacker has the user-authentication information, he or she can easily use a laptop computer to gain access to the wireless LAN.

The best way to counter the rogue AP attack is by constantly scanning for rogue APs in the coverage area for a wireless LAN. Radio scanners can detect the periodic beacon of the APs to determine if there are any rogue APs present in the LAN.

The insertion attacks are also known as intrusion attacks as the intruder, in this case, can easily gain access to the LAN. It is important that a good wireless LAN security policy contains primitives for detecting insertion attacks.

Hijacking Secure Socket Layer (SSL) Connections

Today, Web servers on the Internet use an encryption protocol called Secure Socket Layer (SSL) for secure data transmission over the Internet. Most financial transactions that take place over the Internet, for example stock purchases from an online stockbroker or a book purchase from an online bookseller, take place using the SSL protocol. If a Web server is connected to a wireless LAN and an intruder gets access the wireless LAN, he or she can gain access to the Web server and conduct an attack known as SSL highjacking in which an intruder gains access to the Web server and controls the data.

AP Configuration Parameters

Most APs out of the box from the factory are configured in the least secure mode possible. Adding the proper security configuration is left up to the individual setting up a wireless LAN using the equipment. For example, most APs come with a default SSID. An attacker can use these default SSIDs to attempt to penetrate base stations that are still in their default configuration. Table 12.1 shows some of the most popular APs and their default SSIDs.

Unless the administrator of the APs understands the security risks, most of the base stations will remain at a high-risk level. A good security policy must require that the AP configuration parameters are frequently checked to ensure their proper configuration.

Client Side Configuration Risks

If wireless LAN client computers are incorrectly configured, for example if the security parameters are incorrectly configured or are modified by the user as a mistake, the client computer may reveal critical information that can be picked up by a hacker resulting in the LAN compromise. A good security policy will require that only authorized users modify the client's wireless LAN configuration.

Table 12.1 Popular APs and Their Default SSIDs

MANUFACTURER	SSIDS
Cisco Corporation	tsunami
3Com Corporation	101
Compaq Computer Corporation	Compaq
Intel Corporation	intel
Linksys Corporation	linksys
NetGear Corporation	Wireless

War Driving

War driving is a new activity in which hackers drive around town with a laptop computer equipped with a wireless LAN adapter and a wireless LAN signal monitoring software with the objective of locating APs and recording the GPS coordinates of the AP location. Hackers normally share maps describing the geographic locations of APs on the Internet. If a company has its AP location and information shared on the Internet, its AP becomes a potential target and increases its risk. One of the popular places to upload war driving AP maps is http://www.netstumbler.com. It includes a visual map and a database query tool for locating various APs.

A good security policy will include frequent monitoring of such Web sites and periodic change of the SSIDs of the APs.

Creating Security Policy

A carefully created wireless LAN security policy includes primitives to address most of the security requirements. Creating a security policy for a wireless LAN involves understanding your needs, following a guideline that helps you define the basic parameters that your wireless LAN security policy will enforce, and finally documenting them in an easy-to-follow document that outlines the overall security policy. In this section, we first walk you through a basic guideline that will help you create a security policy; then we show you a sample security policy that can be used as a seed document for your wireless LAN security policy document.

Wireless LAN Security Policy Guidelines

The wireless LAN security policy guidelines vary for each deployment. Following are some of the basic wireless LAN security policy guidelines that can be used to create a security policy for wireless LAN access and management.

Treat All Wireless LAN Devices as Untrusted on Your Network

You should consider all wireless LAN client computers to be untrusted, which means that you assume that any wireless LAN client equipment operating in a LAN could be a rogue computer unless authenticated. Using this primary assumption reminds you not to rely on the inadequate security primitives that many insecure wireless LANs rely upon. For example, if you consider all client computers equipped with wireless LAN adapters as insecure, you will not use MAC address-based authentication as the sole authentication mechanism.

Require the Highest Level of Wireless LAN Authentication You Can Afford

The cost of wireless LAN security infrastructure is falling with advancements in wireless LAN technology. You should try to acquire the highest level of wireless LAN security infrastructure you can afford. You should require in your policy that all APs and client computers must be configured to use the authentication system that is defined in your security policy. For example, use 802.1X authentication protocol for authenticating your wireless LAN users.

Define a Standard Configuration for APs and Wireless LAN Adapters

Your wireless LAN policy must define a standard configuration for wireless LAN adapters and APs. Users deviating from the standard configuration must not be allowed to access the wireless LAN.

Allow Only Authorized Equipment to Be Used in the Wireless LAN

A well-defined security policy will not allow individuals to select their own wireless LAN equipment or software. Though this restriction seems too stiff sometimes, it helps limit the vulnerabilities that unknown equipment may add to the wireless LAN. For example, your policy should allow only a given set of wireless LAN adapters to be used in a wireless LAN.

Discourage Users from Sharing Their Wireless LAN Computers with Unknown Individuals

You should discourage your wireless LAN users from sharing their computers with outsiders. This policy helps keep your wireless LAN configuration information private, available to the LAN users only.

Use Firewalls and VPNs to Secure Your Wireless LAN

Your policy should require that all computers that require high security be protected using firewalls, and all remote access to the LAN must be protected using VPNs.

Enable Strong Encryption When Available

Your policy should choose the strongest available encryption technology and require that all wireless LAN devices use the chosen encryption technology. For example, 802.11 standard uses RC4 as its encryption algorithm and WEP as its security protocol. You should require the use of WEP by all devices that use your wireless LAN.

Allow Only Authorized Personnel Access to APs and Other Critical LAN Equipment

Your wireless LAN security policy must restrict who can manage the LAN equipment. For example, passwords to the AP configuration software must only be distributed among the administrators of the wireless LAN.

Wireless LAN Security Policy at Bonanza Corporation: A Sample Policy

Let's look at the implementation of a wireless LAN security policy in action. The following example involves a technology corporation called Bonanza Corporation. This example is intended to provide you with a general idea that you can use to construct a security policy that may be suitable for your information security needs.

BONANZA CORPORATION

Wireless LAN Security Policy
Attention: All Wireless LAN Users
Policy Effective: Immediately
Today's Date: January 1, 2002

1.0 PURPOSE

This policy establishes information security requirements for Bonanza Corporation offices to ensure that Bonanza Corporation confidential information and technologies are not compromised, and that production services and other Bonanza Corporation interests are protected.

2.0 SCOPE

This policy applies to all internally connected offices, Bonanza Corporation employees, and third parties who access Bonanza Corporation's offices. All existing and future equipment, which fall under the scope of this policy, must be configured according to the referenced documents. DMZ servers and stand-alone computers are exempt from this policy. However, DMZ computers must comply with the DMZ Security Policy.

3.0 POLICY

   3.1 Ownership Responsibilities

1. All office managers are responsible for providing headquarters IT manager, a point of contact (POC), and a backup POC for each office. Office owners must maintain up-to-date POC information with IT and the Corporate Enterprise Management Team. Office managers or their backups must be available around the clock for emergencies, otherwise actions will be taken without their involvement.

2. Office managers are responsible for the security of their offices and the offices' impact on the corporate production network and any other networks. Office managers are responsible for adherence to this policy and associated processes. Where policies and procedures are undefined, office managers must do their best to safeguard Bonanza Corporation from security vulnerabilities.

(continues)

BONANZA CORPORATION *(Continued)*

3. Office Managers are responsible for the office's compliance with all Bonanza Corporation wireless LAN security policies. The following are particularly important: Password Policy for networking devices and hosts, Wireless Security Policy, Anti-Virus Policy, and physical security.

4. The Office Manager is responsible for controlling office access. Access to any given office will only be granted by the office manager or designee to those individuals with an immediate business need within the office, either short term or as defined by their ongoing job function. This includes continually monitoring the access list to ensure that those who no longer require access to the office have their access terminated.

5. The Network Support Organization must maintain a firewall device between the corporate production network and all office equipment.

6. The Network Support Organization and/or SecCommittee reserve the right to interrupt office connections that impact the corporate production network negatively or pose a security risk.

7. The Network Support Organization must record all office IP addresses, which are routed within Bonanza Corporation networks, in Enterprise Address Management databases along with current contact information for that office.

8. Any office that wants to add an external connection must provide a diagram and documentation to SecCommittee with business justification, the equipment, and the IP address space information. SecCommittee will review for security concerns and must approve before such connections are implemented.

9. All user passwords must comply with Bonanza Corporation's Password Policy. In addition, individual user accounts on any office device must be deleted when no longer authorized within three (3) days. Group account passwords on office computers (Unix, Windows, and so on) must be changed quarterly (once every 3 months). For any office device that contains Bonanza Corporation proprietary information, group account passwords must be changed within three (3) days following a change in group membership.

10. No office shall provide production services. Production services are defined as ongoing and shared business critical services that generate revenue streams or provide customer capabilities. These should be managed by a <proper support> organization.

11. SecCommittee will address noncompliance waiver requests on a case-by-case basis and approve waivers if justified.

BONANZA CORPORATION *(Continued)*

3.2 General Configuration Requirements

1. All traffic between the corporate production and the office network must go through a Network–Support–Organization-maintained firewall. Office network devices (including wireless) must not cross-connect the office and production networks.

2. Original firewall configurations and any changes thereto must be reviewed and approved by SecCommittee. SecCommittee may require security improvements as needed.

3. Offices are prohibited from engaging in port scanning, network auto-discovery, traffic spamming/flooding, and other similar activities that negatively impact the corporate network and/or non-Bonanza Corporation networks. These activities must be restricted within the office.

4. Traffic between production networks and office networks, as well as traffic between separate office networks, are permitted based on business needs and as long as the traffic does not negatively impact on other networks. Offices must not advertise network services that may compromise production network services or put office confidential information at risk.

5. SecCommittee reserves the right to audit all office-related data and administration processes at any time, including but not limited to inbound and outbound packets, firewalls, and network peripherals.

6. Office-owned gateway devices are required to comply with all Bonanza Corporation product security advisories and must authenticate against the Corporate Authentication servers.

7. The enable password for all office-owned gateway devices must be different from all other equipment passwords in the office. The password must be in accordance with Bonanza Corporation's Password Policy. The password will only be provided to those who are authorized to administer the office network.

8. In offices where non-Bonanza Corporation personnel have physical access (for example, training offices), direct connectivity to the corporate production network is not allowed. Additionally, no Bonanza Corporation confidential information can reside on any computer equipment in these offices. Connectivity for authorized personnel from these offices can be allowed to the corporate production network only if authenticated against the Corporate Authentication servers, temporary access lists (lock and key), SSH, client VPNs, or similar technology approved by SecCommittee.

9. Infrastructure devices (for example, IP Phones) needing corporate network connectivity must adhere to the Open Areas Policy.

(continues)

BONANZA CORPORATION *(Continued)*

10. All office external connection requests must be reviewed and approved by SecCommittee. Analog or ISDN lines must be configured to accept only trusted call numbers. Strong passwords must be used for authentication.

11. All office networks with external connections must not be connected to Bonanza Corporation corporate production network or any other internal network directly or via a wireless connection, or via any other form of computing equipment. A waiver from SecCommittee is required where air-gapping is not possible (for example, Partner Connections to third–party networks).

4.0 ENFORCEMENT

Any employee found to have violated this policy may be subject to disciplinary action, up to and including termination of employment.

5.0 DEFINITIONS

Internal. An office that is within Bonanza Corporation's corporate firewall and connected to Bonanza Corporation's corporate production network.

SecCommittee. The Bonanza IT Security committee that prepared this document.

Network Support Organization. Any SecCommittee-approved Bonanza Corporation support organization that manages the networking of nonoffice networks.

Office Manager. The individual responsible for all office activities and personnel.

Office. An Office is any nonproduction environment, intended specifically for developing, demonstrating, training, and/or testing of a product.

External Connections (also known as DMZ). External connections include (but are not limited to) third-party data network-to-network, analog and ISDN data lines, or any other Telco data lines.

Office-Owned Gateway Device. An office-owned gateway device is the office device that connects the office network to the rest of Bonanza Corporation network. All traffic between the office and the corporate production network must pass through the office-owned gateway device unless approved by SecCommittee.

Telco. A Telco is the equivalent to a service provider. Telcos offer network connectivity, for example, T1, T3, OC3, OC12, or DSL. Telcos are sometimes referred to as "baby bells," although Sprint and AT&T are also considered Telcos. Telco interfaces include BRI, or:

Basic Rate Interface. A structure commonly used for ISDN service, and PRI (Primary Rate Interface).

> **BONANZA CORPORATION *(Continued)***
>
> **Primary Rate Interface.** A structure for voice/dial-up service.
>
> **Traffic.** Mass volume of unauthorized and/or unsolicited network spamming/flooding traffic.
>
> **Firewall.** A device that controls access between networks. It can be a PIX, a router with access control lists, or similar security devices approved by SecCommittee.
>
> **Extranet.** Connections between third parties that require access to connections of nonpublic Bonanza Corporation resources, as defined in SecCommittee's Extranet policy (link).
>
> **DMZ (Demilitarized Zone).** This describes the network that exists outside of primary corporate firewalls, but are still under Bonanza Corporation administrative control.

Communicating Security Policy

The wireless LAN security policy should be added to every organization's compliance policy that uses wireless LANs. The wireless LAN security policy should be briefed to all employees, especially those who will be using the wireless LAN. The policy and its importance should be properly explained to each individual LAN user. The policy document should be placed along with other corporate documents that define the corporate policies.

Security Policy Compliance

Compiling a wireless LAN security policy and communicating it to users could be a simpler task when compared to ensuring user-compliance. To make sure that wireless LAN users are, in fact, following the security policy, you must monitor their security policy behavior. In addition, any legal policy must be consulted with legal professionals and local law enforcement authorities. Following are some of the commonly practiced ways to monitor security policy in an organization.

- Use computer system logs to ensure that users are following the security policy that you have enforced.
- Make sure that all users frequently change their passwords.
- Users must be required to regularly scan their computers for computer viruses.

Intrusion Detection and Containment

It is important to detect any activity aiming to intrude into the privacy and security of the wireless LAN. All such intrusion activities must be properly detected and contained. Following are some of the common means of detecting intrusion.

Wireless LAN AP Monitoring Software

Wireless LAN AP monitoring software can be used to monitor the presence of APs within a wireless LAN coverage area. Monitoring the APs in a wireless LAN at a given time shows all APs that will be operating at the given time. A rogue AP or an unknown AP operating in a wireless LAN can be easily detected using the monitoring software. If an unauthorized AP is found to be operating within the area that the organization physically controls, it should be immediately turned off and reasons for its operation must be sought from the operators of the AP. If the questionable AP is found to be present in the physical area outside the organization's control, the operators should be contacted to find out whether they are using it for legitimate purposes or the AP belongs to a hacker. If the AP is found to be operated by an unknown entity, law enforcement authorities should be contacted and any possible network security breaches must be assessed.

Intrusion Detection Software

Intrusion detection software operates by constantly monitoring network traffic and activities. Most intrusion detection software is capable of analyzing the network traffic to heuristically determine any known network security breaches and alarm the network administrator (by paging, for example) when they encounter such activities. All intrusion activities must be taken seriously and, if any such activity is found to have happened, all possible security attacks must be properly responded to.

Antivirus Software

Viruses are most common danger to any LAN and standalone computers. Antivirus software can be scheduled to perform routine checks of all network file systems and user computers to make sure that they do not contain files with viruses. Most popular antivirus software, for example Norton Anti-Virus from Symantec Corporation, is updated by manufacturers on a regular basis to provide security from any new computer viruses found.

Firewall and Router Logs

Most firewalls and routers are capable of logging any suspicious activities that could be geared towards destroying, damaging, or degrading a LAN performance or gaining illegal or unauthorized access. For example, most firewalls today are able to deter any denial-of-service (DoS) attacks. They log all network activity that could result in DoS. If a firewall or router log displays any suspicious activity from a computer inside or outside the organization's control, appropriate measures must be taken to deter and or stop such attacks, and law enforcement authorities should be contacted if the threat is of a serious nature.

Network Login and Activity Logs

Most operating systems and authentication servers, for example RADIUS servers, are capable of logging any suspicious login attempt. Hackers, for example, conduct an attack commonly known as the brute-force password attack in which they try to log in to a LAN by attempting possible combinations of username and passwords until they are successful. Attacks of this nature can be easily detected by monitoring these logs frequently.

Getting Ready for Future Security Challenges

While new security techniques are constantly being invented and improved upon, hackers are also busy creating new security threats to LANs and computers in general. Though wireless LANs are a relatively new type of LAN and fewer attacks and threats on wireless LANs are known at this time, it is important to watch out for any new security threats that might become prevalent. To ensure wireless LAN security, it is important that you plan for dealing with the future security challenges by keeping up with the latest development in the security infrastructure of wireless LAN technologies. The use of digital certificates and the public key infrastructure (PKI), for example, must be considered in the near future to provide user authentication and data privacy. Network authentication may also be improved by using newer technologies like DNA fingerprints.

Summary

After deploying a secure wireless LAN, you must continually take measures to ensure long-term LAN security. Establishing and enforcing a wireless LAN security policy helps ensure that staff managing the wireless LAN and the users of the LAN are aware of their responsibilities and roles with regard to a wireless LAN. To successfully establish a wireless LAN security policy that works, you must understand your wireless LAN security requirements, compile a security policy by following a set of guidelines that satisfy your security needs, and communicate the security policy with all wireless LAN users and administrators. In addition to establishing a security policy, you must constantly monitor the policy adherence by the users. You must also set up your LAN to properly detect all intrusion attempts and security breaches. All security breaches must be taken seriously and must be appropriately responded to.

In Appendix A, we will discuss some real-life case studies that show wireless LAN usage in various scenarios. Reading these examples may provide you with a general idea about the feasibility of wireless LANs in your deployment scenario.

APPENDIX A

Wireless LAN Case Studies

Over the last few years, wireless LANs have gained strong popularity among home, SoHo, and enterprise network users. Wireless connectivity of computing devices is rapidly becoming ubiquitous and soon may be, if not the only, certainly the primary method for many portable devices to connect with computer networks. First available at airport kiosks, public access has spread through airport waiting rooms, hotels, and restaurants into coffee shops, hospitals, libraries, schools, and other locations.

In this final part of the book, we examine four case studies that present you with real-life solutions that were implemented to solve networking-related problems. The individual case studies are based on a home wireless LAN, a small corporation wireless LAN, a campus-wide wireless LAN, and a Wireless Internet Service Provider deployment scenario.

> **Home-Based Wireless LAN: The Khwaja Family's House.** In this case study, we discuss the wireless LAN at the house of one of the authors of this book. The case study presents firsthand the experience of setting up a wireless LAN in a century-old home where running cable through the wall could be very difficult and the cost of running a network cable could be inhibiting.

- **A Small Corporation Wireless LAN: The Morristown Financial Group.** The case study for Morristown Financial Group covers the problems that a wireless LAN solved at the corporation.
- **Campus-Wide Wireless LAN: Carnegie Mellon University.** This case study discusses the use of wireless LAN technology at the Carnegie Mellon University campus where LAN connectivity is provided to the users roaming about the campus.
- **Wireless Internet Service Providers: M-33 Access.** The case study briefly explains the problem WISPs are trying to solve and how they go about providing high-speed Internet access over the wireless link.

We hope that the case studies will help you better understand the general wireless LAN deployment issues and the problems they can solve. Let's get started with the case studies.

Home-Based Wireless LANs: The Khwaja Family Residence

Wireless LANs at home have been very successful because they add to the usefulness and enjoyment of computing at home and extend the Internet out of the home office to any convenient place in the house. In this case study, we talk about the use of a wireless LAN at the Khwaja family home.

Background

This case study focuses on the wireless LAN that is being used by Anis Khwaja (one of the authors of this book) and his family. The Khwaja family consists of three growing children, as well as Anis and his wife. This is a family of avid computer users with each member having his or her own computer, as well as communal computers for use in the kitchen and cars. Being a computer professional, Anis also has a small office running various servers and test computers. The primary use for the computers at the Khwaja family is Internet access, which family members use to exchange emails with friends and other members of the extended family.

The Problem

When the Khwaja family moved to their turn-of-the-century, unmodified Victorian home in Long Island, Anis started to look around for a means to set up LAN connections among all the computers. He realized that setting up a wired Ethernet LAN would be a nightmare as the house was built in the early 1900s and still has knob and tube wiring and original construction. Given that the entire house was in the original condition, the Khwajas could not bring themselves to drill holes in the walls and make other modifications that could be necessary to run Ethernet wiring throughout the home. The cost of running an Ethernet cable would also have been a factor to consider. In addition to the difficulty associated with the LAN wiring, the Khwajas were also interested in the ability to roam about within the property with their laptop computers. Anis also wanted the flexibility to work on the computer outside during the spring and summer months. Additionally, Anis wanted the ability to update the MP3 music files and GPS data from the Internet on the computer in the car without taking the computer out of his car. He also wanted to share the DSL Internet connection among the computers. Their budget also demanded that their LAN solution be competitively priced.

The Solution

Anis realized that he needed a wireless solution to solve his LAN problems. He was quick to research the wireless LAN equipment market, and he picked up the 802.11b wireless LAN equipment for his home. He initially decided to perform a pilot to ensure that equipment from multiple vendors would be compatible and the signal would be strong enough to provide a high-quality wireless signal throughout the property.

Anis performed research on the Internet to ensure that his knowledge of wireless LAN technology was up-to-date and that IEEE 802.11b was the most affordable of the wireless LAN solutions available on the market. He bought a Linksys wireless AP that came with a cable router. He also purchased two OriNOCO and Cisco Aeronet PC Card-based 802.11b wireless LAN adapters for laptop computers and an Apple AirPort card for his iMac computer.

With the help of his son, Anis installed the AP according to the manufacturer's instructions in the middle of the room near the television where the cable service was already installed. They had to run a single Ethernet cable between the AP and the Ethernet switch located in Anis's office, which connected the other servers and printers to the wireless LAN. He used his laptop computer installed with Windows XP and the wireless LAN adapter to gauge the relative strength of the AP signal. He located the best spot for the AP and correctly adjusted the antennas to make sure that he receives the best signals at locations within the house where his family plans to use the wireless computers the most.

Once Anis was able to successfully connect to the Internet from his laptop through the wireless link through the AP, he was quite happy to see the results considering that now he would be able to work on his computer outside on the patio while enjoying the garden during the summer.

Results

Wireless connectivity combined with the "always-on" Internet connectivity has had a dramatic effect on the Khwaja family lifestyle within a few short weeks. Now they are able to look for information on the Internet while watching television when anything piques their interest. They also have a voicemail service by http://buzme.com, which has software that notifies the family for any incoming voicemails over the Internet. Anis has interfaced several X-10-based home automation systems with the wireless LAN, which turn on a couple of lights whenever a new voicemail is received by buzme.com.

Anis ended up connecting all computers at home with the wireless connectivity. He thinks he did not just provide his family with instant Internet access throughout the home but also saved tons of money by using wireless LANs compared to wired LAN.

Future

Looking forward, Anis plans to set up a 802.11a-based 54 Mbps wireless link between the computer in the car and a new video server that he is building to be able to upload movies for the road to the computer in the family car.

A Small Corporation Wireless LAN: The Morristown Financial Group

Wireless LANs allow small corporations the capability to construct LANs at a fraction of the price compared to a wired LAN. In addition to the cost, wireless LANs are not only easy to deploy but require less management compared to wired LANs. In this case study, we discuss the wireless LAN at the Morristown Financial Group.

Background

Morristown financial group is a New Jersey-based financial planning and services firm that uses a mix of older WaveLAN and new ORiNOCO equipment from Agere Systems for its wireless network. When they started expanding five years ago, managing partner John Hyland faced a turning point. Adding new employees meant the company finally needed to invest in a networking system. Until then, the company had consisted of only six employees, and setting up a network had not been a priority. But with a growing workforce, sharing client and data management software by sneakernet was no longer practical.

This case study discusses the problem that Morristown Financial Group faced with regards to the wireless LAN and the solution they used to address the LAN issues.

The Problem

The firm started expanding five years ago, and managing partner John Hyland faced a substantial capital expense decision. If he hired the additional employees that the firm needed, he would also need a networking system so that the additional employees could easily share documents. But the firm was going to be relocating within the year, and he didn't want to pay twice for network installation and cabling. "We were growing rapidly and knew we needed to relocate. We didn't want to spend time and money wiring the office space and then three months later abandon it. So the wireless solution was perfect because we could just unplug it and take it with us to the new space."

The Solution

The firm originally selected the WavePOINT II (now called AP1000) access point because its dual-slot architecture could support twice as many users as single-slot access points and also provides an easy migration path to new wireless technologies as they develop.

In 1998, Hyland hired a New Jersey company called InvisiNet to install the workstations. The first system was only a 2 MB system. In 1999, the firm upgraded to 11 MBs. After InvisiNet installed the access point and the network cards in the various computers and laptops, Hyland said that implementing the new system was very fast, simple, and problem-free, and the staff easily adjusted.

Hyland decided to fully integrate the wireless network into his office. One hundred percent of his employees—all 14 workstations—are on the ORiNOCO wireless network. Their DSL Internet connects to an Efficient Networks Flow point hub. In turn, that connects to the ORiNOCO wireless access point. There are no cables connecting the various desktop and laptop computers to the access point — his network is totally wireless.

The Results

According to E.J. von Schaumburg, InvisiNet's CEO, "Morristown Financial Group has moved four times, and all Hyland's had to do is turn off the computers, move them, then turn them back on, and the network is up and ready to go. It's great because he doesn't have to wire the office—an office he does not own! He literally has one piece of wire that goes from the DSL to the access point."

The Future

Morristown Financial Group sees many opportunities cropping up that will utilize their wireless connectivity. For one, the firm has found that the wireless technology (Agere Systems' ORiNOCO products) has helped them to wow their clients. "It's a nice feature to be able to sit at the conference room table with wireless laptops connected to the Internet and show clients immediately how to remotely access their accounts." According to Hyland, "It's much more impressive than pulling them into my office and turning on my desktop. It's a 'wow' factor. We work with a lot of clients who are a little older, so this technology is really impressive to them."

Hyland says that the firm is also able to work with a neighboring accounting firm with much more ease than a wired network would have allowed. "We're doing some partnering with a CPA firm that is on the same floor as we are, and the wireless solution allows us to share files easily on the network. The CPA firm will be buying the technology for one or two workstations so that our two companies can work together on this project easily." Certain files on the network are protected so that only staff with a password can access them. Other files are accessible to anyone on the network.

Campus-Wide Wireless LAN: Carnegie Mellon University

Wireless LANs are quickly becoming very attractive to academic institutions where a college, university, or school of higher learning needs to provide LAN and Internet connectivity to students who are constantly on the move between classes, labs, and various locations within the campus. In this case study, we talk about one of the biggest campus-wide wireless LANs that is in use at the Carnegie Mellon University.

Background

In 1900 industrialist Andrew Carnegie founded Carnegie Mellon University (CMU) as Carnegie Technical Schools in the city of Pittsburgh, Pennsylvania. Today, the university is made up of seven colleges and numerous world-famous research institutes. Carnegie Mellon is a national research university of about 7,500 students and 3,000 faculty, research, and administrative staff.

CMU has always been on the cutting edge of technology. An example of this quick technology adoption and innovation was the introduction of the university's "Andrew" computing network in the mid-1980s. This pioneering network, which linked all computers and workstations on campus, set the standard for educational computing and firmly established the university as a leader in the uses of technology in education and research. When wireless computing started taking off, CMU did not want to stay behind. The staff at CMU looked at their needs and quickly came up with a plan to deploy the campus-wide wireless LAN, which the staff at CMU still claims to be the biggest campus-based wireless LAN anywhere.

In this case study, we talk about the problem that wireless LANs solved at CMU, the solution that was deployed, the results achieved, and the future as seen by the experts at CMU.

The Problem

With growing numbers of mobile computers used by both students and faculty members at the campus, the Andrew network faced problems that related to providing instant Internet access to the mobile computers inside the classroom and within the open areas in the campus. It was envisioned by the computing services staff that sooner or later each student and faculty member would be carrying a computing device that would need access to the Internet and to the wired LAN at the CMU. The computing services staff started looking for solutions.

The Solution

The solution found by the computing staff was to establish a wireless LAN across the campus. Wireless Andrew, the high-speed wireless infrastructure installed at Carnegie Mellon University, is the largest installation of its type anywhere. Started as a research network in 1994 to support Carnegie Mellon's wireless research initiative, Wireless Andrew has been dramatically expanded since its conception. Wireless Andrew has been installed in many of the academic and administrative buildings and will soon cover all academic and administrative buildings on the Carnegie Mellon campus.

The project started in August 1994 with an award of $550,000 from National Science Foundation (NSF) for the two-year project to construct a campus-wide wireless LAN. Phase One of Wireless Andrew began in February 1997 and provided services such as file transfer, email, and access to the library and databases along with complete Internet services. The eventual design was to install approximately 200 access points throughout 12 buildings on the campus. One of the objectives of this project was to support research and development of mobile and nomadic computing. All of the university research programs concerning wireless computing came to be known generally as the Wireless Initiative. Phase One of Wireless Andrew was released in 1997.

The Initial Wireless Andrew

Wireless Andrew consists of wireless LAN access points that are connected to the wired Andrew Network to provide seamless connectivity between the wired and the wireless world. The technology used in the Wireless Andrew project contains only a few more components than you would normally find in a peer-to-peer wireless LAN (see Chapter 2 for more information on peer-to-peer wireless LANs).

For the Wireless Andrew project the PC Card used in laptop computers was the WaveLAN PCMIA wireless LAN interface card along with the cellular digital packet modems (CDPD). The CDPD service permits roaming outside the campus network throughout the Pittsburgh area. CDPD also supports Internet Protocols (IP) that allow the CDPD network to be linked with the WaveLAN network. CDPD uses the idle channels in the cellular system to provide the connectionless digital packet service (CDPD). The coverage area for the CDPD system has a radius of 1-10 miles. When added to the WaveLAN network infrastructure, it provides additional roaming capability for the campus wide network.

Phase Two of Wireless Andrew was deployed in August 2000.

The New Wireless Andrew

In August of 2000, Wireless Andrew was upgraded to use the new IEEE 802.11b DSSS-based technology. The upgrade more than tripled the throughput of the network and brought the client software up to the supported Lucent version. The network meets the 2.4-GHz, IEEE 802.11b Direct Sequence Spread Spectrum wireless Ethernet standard. The upgrade also included connectivity to all 32 academic and administrative buildings and key outdoor areas located on the main campus. The upgrade also eliminated the previous 915-MHz wireless research network that was disconnected and its users converted to the production network.

The Challenges

One of the challenges in laying out a multiple access point wireless LAN installation was planning the layout of the access points and ensuring that adequate radio coverage is provided throughout the service area. The experience of the computing staff at CMU told them that the layout must be based on measurements, not just on rule of thumb calculations. These measurements involved extensive testing and careful consideration of radio propagation issues when the service area was large, for example an entire campus in CMU installation. Even a very carefully considered access point layout had to be modified after installation was complete in order to remedy coverage gaps.

The other big challenge was the need to provide the best possible seamless roaming capability to the wireless LAN users. The wireless LAN deployment team achieved this goal by carefully planning the access point sites such that signals from various APs overlapped each other, resulting in minimal lapse in connectivity when a wireless LAN device is in motion.

The Results

The completion of the installation in and around the academic and administrative buildings translates into seamless connectivity for over 1,700 wireless users across campus, in all classrooms, common spaces, offices, and many outdoor areas. The Wireless Andrew service is available to faculty, staff, and students and offers wireless data connections at speeds up to 11 Mbps. Users in all administrative, residential, and academic buildings as well as key outdoor areas located around the main campus can enjoy wireless networking with Lucent's 2.4-GHz, IEEE 802.11b Direct Sequence Spread Spectrum wireless Ethernet standard. No other campus of this size has such complete coverage.

Wireless Internet Service Providers: M-33 Access

Wireless access to the Internet is quickly becoming one of the hottest businesses in the Internet services area. This case study focuses on an wireless Internet service provider, M-33 Access. This case study will provide you with basic ideas relating a wireless ISP setup.

Background

Founded in 1999, M-33 Access (named after main highway M-33 in Rose City, Michigan) offers high-speed wireless Internet and wide area networking (WAN) services to thousands of potential clients over a 20,000 square mile area throughout Northeastern Michigan. Using Outdoor Router equipment from Agere Systems ORiNOCO, M-33 Access has over 20 towers that are providing the only high-speed wireless Internet access to approximately 30 cities in the coverage area. M-33 also uses the outdoor routers to replace direct-leased lines and to connect banks of dial-up modems to remote locations.

The Problem

A software programmer by trade, M-33 founder Glenn Wilson started M-33 Access when he moved to rural Michigan in the winter of 1998. "We had just sold the two ISP services we built because there was no Internet available where we lived in Colorado. We moved to Michigan so that we could spend more time with our family and so I could get back to programming. Just our luck, we again moved to a town that had no available Internet access!" said Wilson. "If you wanted access, you had to dial-out long distance or use the modems at the local library. It was ridiculous—small towns and highly rural areas were being ignored by the large ISP companies while the rest of the world was in touch."

Wilson purchased a system with dial-up modem boards to provide wired, dial-up access to the Internet for residents of Wilson's hometown of Rose City and the neighboring town of Mio (pronounced My-o). This worked fine for the residents of Rose City and Mio, but Wilson ran into problems when he tried to put modems in the neighboring town of Hale. The problem was that Wilson had to work with two different phone companies to lease T1 lines. At the time, the phone companies in the area hadn't established business relationships with each other, and it was very expensive to run T1s from one service area to another.

The Solution

It was at this time that Wilson learned of Agere's ORiNOCO wireless access points, antennas, and amplifiers, and decided to take a plunge into the newest broadband technology. Wilson built and outfitted a 195-foot tower in Hale, as well as one in Rose City. Those towers enabled them to create a wireless point-to-point link that stretched 19 miles between the cities of Hale and Rose City, Michigan. This meant that residents of Hale, Rose City, and other neighboring communities now had high-speed wireless Internet access available, something the telephone companies were unable to affordably provide.

"It's important to provide good service to people," Wilson said. "If your customers don't know who to call for service and they are getting bounced all over the place, they are going to blame the ISP for a bad phone line connection. Our broadband technology in most of our areas connects the Internet directly from us to the customer."

M-33 Access's network service area consists of 30 of Agere's ORiNOCO Outdoor Routers (OR) mounted on about 20 towers. Each Outdoor Router can provide a broadband Internet connection directly to a single user's computer or to a wireless LAN that then distributes the connection throughout a building to all of a business's computers. Each ORiNOCO Central Outdoor Router (COR) can accommodate up to 30 individual wireless clients or WLAN networks. The Outdoor Router can also be used to bridge Internet access to 16 additional outdoor routers.

M-33 Access's customers include county buildings, police and fire departments, schools, businesses, and homes. The wireless Internet access makes life much easier for these customers and, in most cases, eliminates the need for a second or additional phone line or dedicated connections at high cost.

Special mobile solutions are also available. For example, building contractors in M-33's service area are often required to respond to online insurance company requests. When a contractor is on-site at a job, it can be very inconvenient to have to leave the job site, go back to the office to respond to the request, and then return to the job site to finish working. Within the M-33 wireless broadband service area, a contractor can now just flip up an antenna on his truck, turn on his laptop computer, and go online right then and there.

Also, police officers will soon be able to sit in their cars and complete their paperwork on a laptop computer, then upload the completed forms to the station's database via the wireless connection. They will no longer have to go to the station to fill out paperwork.

M-33 also provides ORiNOCO wireless access points for businesses and individuals that need network connectivity in their offices and homes.

Challenges in Installation

The main challenge for new wireless installations lies in positioning the antennas high enough off the ground to provide a good signal. Michigan is full of very tall, dense trees and hills. Another challenge is weather. If cable connections get wet from rain or snow, they do not function properly. Range and signal strength can be diminished substantially if the connections get wet. M-33 has used several methods to keep connections dry, including wrapping them with special tape that acts like tar and seals them. Since the tape freezes in the winter, it is difficult to work with.

To deal with downed tower antennas or amplifiers, M-33 places a hot spare antenna on each tower. This way, in the dead of winter if one antenna goes out during inclement weather, service technicians do not have to climb the tower to replace it. He can simply unplug the "bad" antenna and plug the "good" one in. When the weather is better, they can climb the tower and correct the problem.

Service Usage Billing

All billing goes directly through M-33 Access. M-33 Access charges as little as $75 per month for high-speed wireless broadband for businesses, and $50 per month for homes. Hardware, installation, and setup are extra. Additional networking services are also available and easily accomplished through the wireless routers.

M-33 also offers a Standard Equipment Package for $350. This package includes one antenna, 30 feet of networking cable, cable ends, one PigTail, one PC Card, and one PCMCIA slot (if necessary).

Security

As with any wireless network, security is of concern to M-33 Access and its users. To provide wireless security, M-33 encourages its customers to activate Wired Equivalent Privacy (WEP) encryption, the basic level of wireless LAN security assigned to the 802.11b wireless LAN specification.

"Most people don't activate WEP automatically, so it's important to educate them to do so," said Wilson. "Otherwise, the possibility exists that an intelligent hacker could intercept a customer's data."

The company also encourages its customers to use virtual private networking (VPN) applications. VPN allows data to be encrypted before it gets sent from one wireless access point to the public network. This means that data is encrypted at the sender's network address and decrypted only at the receiving end. VPN can provide an additional layer of network security by encrypting both the sending and receiving network addresses.

The Result

"Things become possible that were only imagined! From cyber cafes, to hotels, to business connectivity, M-33 Access can allow customers to access the Internet and even access documents on their corporate intranet while they have lunch or dinner. Our customers can get on the Internet even when they are out on their pontoon boat fishing! People can also set up wireless cameras to monitor their place of business or home while they are away. There is even a radio station that streams its broadcast over the Internet because of the M-33 wireless network."

In addition to M-33 Access providing wireless Internet, they are also providing wide area networking (WAN) services to rural Michigan as well. For example, there are four newspapers located in M-33's service area. M-33 has enabled the four sites to be networked together via the wireless equipment. The newspapers can now easily transfer news stories back and forth, and share printers and files across a 100-mile distance! Additional services such as VPN, voiceover IP, and many other networking solutions are possible.

M-33 Access's services continued to grow, and today the company can provide high-speed wireless broadband access to thousands of prospective customers over a 20,000 square mile area in northeastern Michigan.

The Future

"The sky really is the limit. We plan to continue expansion here in northeastern Michigan until we hit water. It's the people in rural areas that get left behind when it comes to information technology," Wilson said. "It's important to keep these people connected and provide them with easy, affordable access to the Internet. I don't plan on stopping until people in rural areas everywhere have Internet access at their fingertips."

APPENDIX B

Installing ORiNOCO PC Card Under Various Operating Systems

In Chapter 9, "Equipment Provisioning and LAN Setup," we explained the procedure for setting up the ORiNOCO wireless LAN adapter under Windows XP. In the paragraphs that follow, we discuss the installation process of ORiNOCO PC Card for the following operating systems:

- Windows 98, Windows ME, and Windows 2000
- Windows NT
- Mac OS
- Linux

Installing under Windows 98, Windows ME, and Windows 2000

Windows 98, Windows ME, and Windows 2000 are called plug-and-play operating systems because they contain software that automatically detects the addition and removal of peripheral devices and can help you set up hardware devices in an easier way compared to Windows NT, which does not support plug-and-play technology.

All Microsoft Windows 2000 operating systems include certified drivers for your ORiNOCO PC Card. Although you can use these drivers to install your PC Card, we advise you to use the driver provided on the ORiNOCO CD-ROM. The new driver is easier to use, includes online Help, and allows you to create multiple network profiles that you can use for different configurations.

The setup procedures for Windows 98, Windows ME, and Windows 2000 are very similar. The procedure that you will need to follow to correctly install and set up ORiNOCO PC Card wireless LAN adapter follows.

System Requirements

Following are the system requirements for installing an ORiNOCO Gold/Silver PC Card under Windows 98, Windows ME, and Windows 2000.

- An empty PC Card or CardBus Slot.
- Administrative privileges for the computer you are about to install the ORiNOCO PC Card in.
- An ORiNOCO PC Card wireless LAN adapter.

Software Requirements

You must have the following software.

- The ORiNOCO CD-ROM that was included with your PC Card kit, or
- Drivers installed in a directory on your computer hard disk if you manually downloaded them from the ORiNOCO Web site.

Installation Steps

The complete installation of your PC Card and the necessary drivers (for Windows 98, Windows ME, and Windows 2000) consist of the following steps:

1. Install the ORiNOCO PC Card software drivers.
2. Set basic parameters.
3. Finish installation.

This section describes the installation of the ORiNOCO PC Card for Microsoft Windows 98, Windows ME, and Windows 2000 operating systems only.

Installing ORiNOCO PC Card Under Various Operating Systems 289

Installing ORiNOCO PC Card Software Drivers

Windows 98, Windows ME, and Windows 2000 operating systems support Plug and Play for PC Cards. Once you insert the ORiNOCO PC Card into your computer, these operating systems will automatically detect the card, and enable the ORiNOCO Driver, or it will start the Add New Hardware wizard and prompt you to install the driver, when the operating system cannot find the required driver.

This would typically occur when inserting the ORiNOCO PC Card into your computer for the very first time. To install the driver, proceed as follows:

1. If Windows starts the Add New Hardware wizard, follow the instructions of the New Hardware Found wizard to install the drivers. Then you will be prompted to locate the driver installation files.

2. Select the ORiNOCO CD-ROM that was included with your PC Card kit and, if you downloaded the drivers from the ORiNOCO Web site, navigate to the folder that matches your Operating system. For example, if the folder on your hard disk where you installed the downloaded driver matches D:\Drivers or if your hard disk driver letter was D then, depending on your operating system, you will use one of these directories:

   Windows 98: D:\Drivers\Win_98.

   Windows ME: D:\Drivers\Win_ME.

   Windows 2000: D:\Drivers\Win_2000.

   When finished installing the drivers, Windows automatically opens the Add/Edit Configuration Profile window. The Add/Edit Configuration Profiles let you set up the basic radio profiles that you can use with your wireless LAN adapter.

Setting up the Configuration Profiles

After installing the drivers, Windows will open the Add/Edit Configuration Profile window for your ORiNOCO PC Card as pictured in Figure B.1.

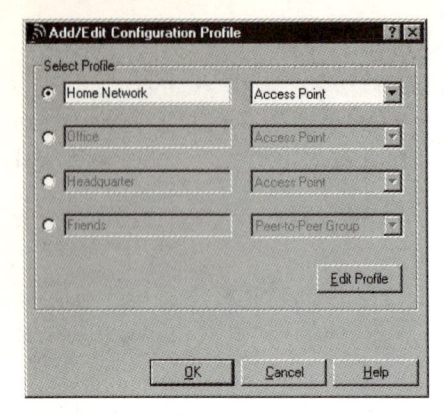

Figure B.1 Add/Edit Configuration Profile window.

The Add/Edit Configuration Profile window enables you to specify one or more network connection profiles. For example, you can set up profiles for an office, to connect to an enterprise network via an AP, or for workgroup computing to share files with colleagues or friends in small peer-to-pcer workgroups without an access point. You may also configure a profile for home, to connect to a residential gateway (RG) that provides access to the Internet or your home printers. The ORiNOCO Edit Configuration window also provides you the possibility to change other parameters (Encryption, Advanced, and Admin parameters). You are advised to leave these parameters to their default settings unless there are special situations, for example, upon advice of an ORiNOCO expert. To connect your computer to a wireless network you will need to set up the following values:

1. Assign a name to the network connection profile.
2. Use the pull-down menu on the right to select how you wish to connect to the wireless network.
3. Click the Edit Profile button to view/modify the parameters for the selected profile.

For first-time installations, you are advised to set up the single profile using only the basic settings. The Basic Settings window is shown in Figure B.2.

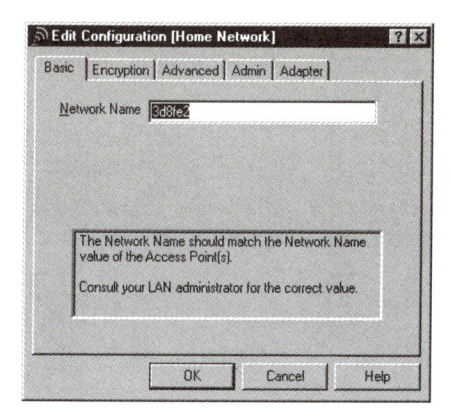

Figure B.2 Basic Settings tab in Edit Configuration window.

Basic Settings for Enterprise Networks

If you wish to connect to an enterprise network, use the Add/Edit Configuration Profile window select to connect to an access point and set the correct network name or SSID:

1. In the field Network Name, define the name of the wireless network to which you want to connect. You can use either the value "ANY" to connect to any wireless network in the vicinity of your computer, or an "exact" value to connect to a known wireless network. Consult your LAN administrator for the value that applies to your network.
2. Click OK to confirm and return to the Add/Edit Configuration Profile window.
3. Click OK again to continue with the last steps of the installation as described in the next section, *Finishing the Installation*.

Basic Settings for Using the ORiNOCO PC Card with a Residential Gateway

If you wish to connect to a home network via a residential gateway, use the Add/Edit Configuration Profile window to connect to a residential gateway. Setting up the PC Card for connecting with a residential gateway involves setting up the correct network name and the encryption key. Figure B.3 shows the Encryption Settings window.

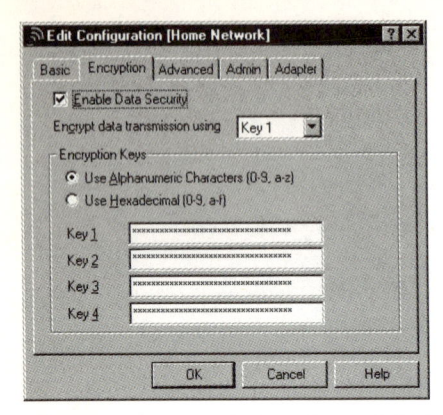

Figure B.3 Encryption Settings tab in Edit Configuration window.

In the field Network Name enter the six-character RG ID to define the name of the wireless network to which you want to connect. The network name has to match the unique RG ID (which can be found on the device).

In the Encryption Key field, enter the last five digits of the RG ID (default).

Note: If you changed the default encryption key on the residential gateway you need to enter the new value here as well.

Click OK to confirm and return to the Add/Edit Configuration Profile window.

Click OK again to continue with the last steps of the installation as described in the next section, *Finishing the Installation.*

Basic Settings for Peer-to-Peer

If you wish to connect to a peer-to-peer workgroup, use the Add/Edit Configuration Profile window to select to connect to a peer-to-peer workgroup. Set the correct network name and encryption key.

1. In the field Network Name, define the name of the wireless network to which you want to connect. If there is already a peer-to-peer group with this name available, your computer will automatically connect to this workgroup. If there is not yet such a group available, your computer will automatically start one with this name.

2. Click OK to confirm and return to the Add/Edit Configuration Profile window.

3. Click OK again to continue with the last steps of the installation as described in the next section, *Finishing the Installation.*

Finishing the Installation

When you have finished with Set Basic Parameters, click the OK button to close the Add/Edit Configuration window and to proceed with the installation process. Windows will finish building the driver configuration database and copy some files to your computer's hard disk. When Windows has finished copying files, it will prompt you to restart your computer. Click the OK button to restart your computer.

After you have restarted your computer, the Windows operating system will detect the ORiNOCO PC Card (you can hear the two-tone sound of the PC Card Socket Controller). Load the ORiNOCO driver, and in the dialog box enter a Windows username and password. The password you enter here will be the one used to login to the Windows Network Neighborhood.

Verifying Installation

Follow the steps below to verify whether the installation of the drivers was completed successfully. Check the LEDs on your PC Card. The following should be visible:

- A steady green Power LED to indicate that the PC Card is active.
- A flickering green Transmit/Receive LED to indicate wireless activity while transmitting data.

Installing Networking Support for the First Time

If this is the very first time that networking support is installed onto your computer, the Windows operating system will prompt you to enter a computer and workgroup name. These names will be used to identify your computer on the Microsoft Network Neighborhood. Follow these steps if you installed the networking support for the first time:

1. In the Computer Name field, enter a unique name for your computer.
2. In the Workgroup field, enter the name of your workgroup.
3. (Optional) Provide a description of the computer in the Computer Description field.

For more information about setting your Windows Network Properties, consult your Windows documentation or the Windows online help information.

Installing under Windows NT 4.0

Installation process for ORiNOCO PC Card under Windows NT is very similar to the previous installation process, except for the following:

1. Windows NT is does not support plug and play the way Windows 98, Windows ME, and Windows 2000 do. Windows NT does not recognize the addition of a hardware device. However, it may recognize removal of a device if the drivers for the device are installed, as the drivers will be unable to locate the hardware device.
2. The PC Card/PCMCIA support may need to be enabled manually.
3. To swap PC Cards, Windows NT machines typically require you to restart the computer in order to recognize the card and load the drivers.

In this section, we will only talk about the unique steps (in addition to those mentioned for installation under Windows 98, Windows ME, and Windows 2000) that you must follow in order to successfully install the ORiNOCO PC Card under Windows NT.

System Requirements

Following are the system requirements for installing the ORiNOCO Gold/Silver PC Card under Windows NT 4.0:

1. An empty PC Card or CardBus Slot.
2. Windows NT administrative privilege for the computer you are about to install the ORiNOCO PC Card in.
3. The ORiNOCO PC Card wireless LAN adapter.

Software Requirements

You must have the following software:

1. The ORiNOCO CD-ROM that was included with your PC Card kit, or
2. Drivers installed in the directory on your computer hard disk if you manually downloaded them from the ORiNOCO Web site.

Installing ORiNOCO PC Card Under Various Operating Systems 295

Installation Steps

Installation of ORiNOCO PC Card in Windows NT includes the following steps:

1. Enabling the PCMCIA Services.
2. Enabling Network Support.
3. Following the steps defined in the previous section, *Installing Under Windows 98, Windows ME, and Windows 2000*.

Enabling PCMCIA Services

If you have not previously enabled the PCMCIA services on your mobile computer to allow the Windows NT operating system to detect PC Cards in the computer's PC Card slot, you must enable the PC Card Socket controller, identified as the PCMCIA device. Follow these steps to enable the PCMCIA service:

1. Click the Start button, then select Settings, and then click Control Panel.
2. Double-click the Devices icon.
3. Scroll down the list of devices and select the item PCMCIA. Figure B.4 shows the PCMCIA service for Windows NT.
4. Click the button Startup on the right side of the Devices window, and set the Startup type for the item PCMCIA to Boot.
5. Click OK to confirm and return to the Devices window.
6. Click Close to return to the Control Panel.

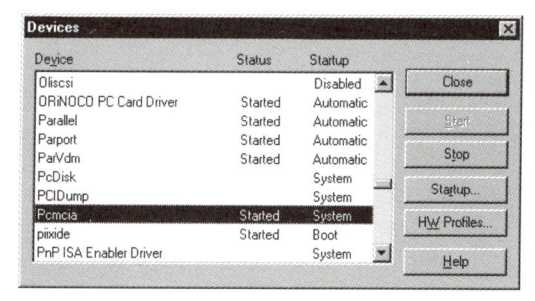

Figure B.4 Enable the PCMCIA service for Windows NT.

Enabling Network Support

To introduce your ORiNOCO network adapter card to the Windows NT operating system, you need to enable Network support for your ORiNOCO wireless station.

1. From the Windows NT Taskbar, click the button.
2. Click Settings and then Control Panel.
3. In the Control Panel window, double-click the Network icon to open the Network Settings window. If no network has been installed yet, you are prompted to install it now. Click Yes to install Windows NT Networking, and follow the instructions as they appear on your screen. If networking support are already installed, you see a window with multiple tabs.
4. Select the tab Adapters, and click the Add button.
5. When prompted to select a driver, select Driver from disks provided by hardware manufacturer, and enter the folder directory where Windows NT can locate the ORiNOCO PC Card drivers. For example, if you are using ORiNOCO CD-ROM to install the software drivers, the applicable folder path may be: d:\drivers\win_nt.
6. Follow the instructions on your screen and restart your computer when prompted to do so.

Finishing the Installation

Configure your ORiNOCO PC Card by following the steps defined in the previous section, *Installing Under Windows 98, Windows ME, and Windows 2000*.

Installing under Mac OS

To connect your Apple PowerBook to a wireless network, you will need to perform the following steps.

1. Install the ORiNOCO PC Card software.
2. Install one or more networking protocols to allow your ORiNOCO computer to communicate with other wireless and/or wired computers on the network.
3. Power up the computer with the ORiNOCO PC Card.
4. Configure the networking protocols of your MAC operating system to select the ORiNOCO interface for network communication.

System Requirements

Following are the hardware requirements for installing the ORiNOCO Gold/Silver PC Card under Mac OS:

- Powerbook Computer with an empty PC Card or CardBus Slot.
- The ORiNOCO PC Card wireless LAN adapter.

Software Requirements

You must have following software:

- ORiNOCO PC Card drivers for Mac OS.

Installation Steps

Follow the steps below to successfully install the ORiNOCO PC Card drivers under Mac OS.

Installing the Software Drivers

To install the PC Card software:

1. Insert the ORiNOCO CD-ROM into your Apple PowerBook.
2. Double-click the CD-ROM icon on the desktop of your computer, to display the contents of the ORiNOCO CD-ROM.
3. Now double-click the file called ORiNOCO Installer to start the installation program.
4. In the Welcome window, click the button Continue to proceed.
5. From the list of options, select Easy Install.
6. Click the Install button to start the installation. If you have any applications open during the ORiNOCO installation process, you are prompted to close these applications. Click No to abort installation, or Click Continue to proceed and have the MAC operating system close all the open applications.
7. Follow the instructions as they appear on your screen and restart your computer.

Installing Networking Protocols

Subject to the type of networking environment, you should install one or more of the following protocols:

- AppleTalk, most commonly used to connect a number of Apple workstations to a networking environment identified as the "AppleTalk zone" or "AppleShare server."
- TCP/IP to connect to larger network infrastructures, and/or allow connections to the Internet via the network.
- Optionally you can install and enable both networking protocols. For more information about installing protocols, please consult the "balloon help" and User's Manual that came with your Apple PowerBook.

To finish the installation of the ORiNOCO drivers and networking protocols, please restart your Apple PowerBook computer.

Enabling Your ORiNOCO PC Card

To enable your PC Card, insert the ORiNOCO PC Card into your computer. Once your PC Card is inserted, you will see the ORiNOCO icon appear on the desktop of your Apple PowerBook. If this icon is not displayed, verify whether the PC Card is properly inserted into the PC Card slot of your Apple PowerBook.

To enable your ORiNOCO connection, you will need to configure the AppleTalk protocol to use the ORiNOCO PC Card for its network communication.

1. Double-click the ORiNOCO icon on your desktop to open the AppleTalk Parameter window.
2. In the AppleTalk window, open the pull-down menu of the field Connect Via. Select the item ORiNOCO (some versions might still show WaveLAN/ IEEE).
3. Click the button on the top-left side of the window to confirm your changes and close the window.
4. Click Save to save the new configuration.

If you are using TCP/IP as networking protocol, you will need to configure the TCP/IP protocol to use the ORiNOCO PC Card for its network connections. You can access the TCP/IP settings via the option Control Panels in the Apple menu. Please consult the "balloon help" and User's Manual that came with your Apple PowerBook for more information.

Installing ORiNOCO PC Card Under Various Operating Systems

Customize PC Card Settings

When you insert the ORiNOCO PC Card into your Apple PowerBook computer, it will start operating with the following factory-set defaults.

Connect to a Network Infrastructure. Use the ORiNOCO Network Name "ANY" to connect to the first access point that provides a communications quality that is acceptable or better. To view or modify the ORiNOCO parameters, open the ORiNOCO Setup window that is listed under the Apple menu, as an item in the Control Panel.

Basic Parameters

For standard operation of your ORiNOCO PC Card, you will need to set only the following parameters, characterized as the Basic Parameters.

- Type of network to which you wish to connect your wireless computer.
- The ORiNOCO Network Name of the network.

To connect to an infrastructure network:

1. Clear the Ad-hoc Demo Mode tick box.
2. If you selected to connect to an infrastructure network, you need to identify the name of that network. Consult your LAN administrator for the value that applies in your situation. You operate your computer in multiple network environments that are identified by different WaveLAN Network Names. You do not know the ORiNOCO Network Name of the network to which you would like to connect your computer.

To connect to an ad-hoc workgroup of wireless stations:

1. Place a check mark in the Ad-Hoc Demo Mode tick box. In this mode your ORiNOCO PC Card will ignore the ORiNOCO Network Name value and the ORiNOCO access points.
2. Fix the radio channel to operate at its factory-set default channel. This means that your ORiNOCO station can communicate with any other ORiNOCO station within its range, provided that these stations have been equipped with cards that have a matching default radio frequency.

Advanced Parameters

The other parameters available from the ORiNOCO Setup window are advanced parameters that work most efficiently when you leave the settings to these parameters to their factory-set defaults. You may need to modify these parameters only when troubleshooting ORiNOCO performance upon the advice of an ORiNOCO expert, or an ORiNOCO Technical Support representative.

Installing under Linux

ORiNOCO PC Cards are one of the few wireless LAN adapters that support the Linux operating system. This section describes how to install ORiNOCO drivers under the Linux operating system.

System Requirements

Following are the system requirements for installing ORiNOCO Gold/Silver PC Card under Linux:

- This software can be compiled and installed on Intel architecture systems running Linux kernel versions 2.0.x, 2.2.x or 2.4.x.

Software Requirements

You must have the following software and support files.

- This software for the ORiNOCO PC Card is distributed in a compressed archive wavelan2_cs-6.16.tar.gz. If you did not receive the software along with the adapter, you can obtain it from ORiNOCO using its Web site at http://www.orinocowireless.com.
- To compile the software you will need to have the full set of Linux kernel source files installed. Depending on the exact version of the kernel, you will need approximately 150 MB of free disk space. Once compiled, the driver will use approximately 40 KB.

Installation Steps

The driver files for the Linux driver are not ready for direct installation onto any Linux computer. To build and install the driver, you need some expertise on the Linux operating system, in general, and the type and version of the kernel installed on your computer. With this knowledge, you can use the driver source files provided to build your own Linux driver for your specific computer and kernel.

Installing ORiNOCO PC Card Under Various Operating Systems 301

Before you Start

- Determine the type and version of the Linux kernel of your computer, and check whether it meets the system requirements listed above.
- Read the Linux PCMCIA-HOWTO by David Hinds. This document is probably provided on a CD-ROM of your Linux distribution. You can download the latest version from: http://pcmcia-cs.sourceforge.net.
- Please read the section titled *Prerequisites and kernel setup* of the PCMCIA-HOWTO.

Build the ORiNOCO Driver

1. Obtain a copy of the Linux PCMCIA package from a CD-ROM of your Linux distribution or download the latest version from: http://pcmcia-cs.sourceforge.net. For your convenience, the latest ORiNOCO CD-ROM contains a copy of the PCMCIA package in subdirectory: Xtras/Linux/PCMCIA

2. To unpack the Linux PCMCIA package, copy it to the current working directory and type:

   ```
   % tar xzvf pcmcia-cs-3.1.29.tar.gz.
   ```

   Note: If you use the archive supplied on the CD-ROM, use archive name "pc3_1_29.tgz" instead of "pcmcia-cs-3.1.29.tar.gz".

   a. Extract the ORiNOCO distribution archive on top of the Linux PCMCIA package.

   ```
   % cd pcmcia-cs-3.1.29
   % tar xzvf ../wavelan2_cs-6.16.tar.gz
   ```

   Note: If you use the archive supplied on the CD-ROM, use archive name "../wlli616.tgz" instead of "../wavelan2_cs-6.16.tar.gz".

3. To build and install the driver, follow the procedure below.

   ```
   % make config
   ```

   Answer the presented questions. Usually the default answers are OK and pressing "Enter" is enough. On RedHat 7.1 systems, however, you should specify "/usr/src/linux-2.4" as the Linux source directory instead of the default "/usr/src/linux". For more detailed information on configuration, building, and installing, see the PCMCIA-HOWTO as mentioned in the *Before you Start* section. Now run the Build script.

   ```
   % ./Build
   ```

This script determines whether your system uses in-kernel PCMCIA and either builds the full PCMCIA package or just the driver. Before installing the driver with the Install script, you must become "root."

```
% su
..
# ./Install
```

This script determines whether your system uses in-kernel PCMCIA and either installs the full PCMCIA package or just the driver.

Configure Your ORiNOCO PC Card

Before configuring the driver through module parameters (in /etc/pcmcia/config.opts), make sure that /etc/pcmcia/wireless.opts file is either absent or contains blank parameter values as shown below.

```
*,*,*,00:60:1D:*|*,*,*,00:02:2D:*)
    INFO=""
    ESSID=""
    MODE=""
    KEY=""
;;
```

To configure the ORiNOCO PC Card, please refer to the online manual page (wavelan2_cs.4).

```
% man wavelan2_cs
```

- You should also consult the network adapter sections of the PCMCIA documentation.

```
% more PCMCIA-HOWTO
```

Use an editor to configure the module parameters:

```
# vi /etc/pcmcia/config.opts
```

For example:

To connect your computer to a wireless infrastructure that includes access points such as the Agere Systems AP-1000 or AP-500, you will need to identify the network name of the wireless infrastructure. For example if your infrastructure uses the network name "My Network", edit the config.opts file to include the following.

```
module "wavelan2_cs" opts "network_name=My\ Network"
```

Installing ORiNOCO PC Card Under Various Operating Systems 303

Notice that the space character needs to be escaped with a backslash.

To connect your computer to an Agere Systems Residential Gateway RG-1000, you will need to know the RG ID (=network_name) and the encryption key. You can find the RG ID on a small label on the rear of the unit. For example, if your RG-1000 has ID 225ccf and you did not change the encryption key yet, edit the config.opts file to include the following:

```
module "wavelan2_cs" opts "network_name=\"225ccf\" key_1=\"25ccf\"
enable_encryption=Y"
```

If you changed your encryption key, you should specify this key as key_1 on the parameter line.

To connect your computer to a peer-to-peer network, in an environment without access points, the IBSS mode is recommended. For example to connect to a peer-to-peer network called My Network, enter the following in the config.opts file:

```
module "wavelan2_cs" opts "create_ibss=Y network_name=My\ Network"
```

Optionally you can also include a Station Name value that can be used to identify your computer on the wireless network. For example if you wish to name your computer "Wave1" when connecting it to a wireless infrastructure, edit the config.opts file to include the following:

```
module "wavelan2_cs" opts "network_name=Ocean station_name=Wave1"
```

To connect your computer to an ad-hoc workgroup of wireless computers, enter the following in the config.opts file:

```
module "wavelan2_cs" opts "port_type=3"
```

Note that the "Ad-Hoc Demo Mode" is not the recommended mode for a peer-to-peer network. The configuration of this non-interoperable mode is only explained here for special applications (for example, research, or compatibility with other previous WaveLAN/IEEE products).

The IBSS mode described in c) is the preferred and interoperable mode for creating a peer-to-peer network.

Use an editor to modify the network options for your adapter.

```
# vi /etc/pcmcia/network.opts
```

The parameters need to be correct for the connected network. Check with your system administrator for the correct network information. Refer to the PCMCIA-HOWTO for more configuration information.

Appendix B

For example:

```
*,*,*,*)
    IF_PORT=""
    BOOTP="n"
    IPADDR="10.0.0.5"
    NETMASK="255.255.255.0"
    NETWORK="10.0.0.0"
    BROADCAST="10.0.0.255"
    GATEWAY="10.0.0.1"
    DOMAIN="domain.org"
    DNS_1="dns1.domain.org"
    ;;
```

Note that RedHat 7.1 doesn't use the network.opts to configure the driver. Instead it uses a GUI-based tool called "netcfg" that creates scripts, like ifcfg-eth0, in the directory /etc/sysconfig/network-scripts.

Using the default GNOME menu, you can start netcfg from: Programs->System->Network Configuration

Restart the PCMCIA services.

```
# /etc/rc.d/rc.pcmcia restart
```

or

```
# /etc/rc.d/init.d/pcmcia restart
```

For a more detailed description about the various configuration options and definitions, please consult the ORiNOCO documentation in Adobe's Acrobat PDF format on the CD-ROM that came with your product.

ORiNOCO User documentation is also available on the ORiNOCO Web site: http://www.orinocowireless.com.

Glossary of Terms and Abbreviations

Access Control List (ACL)   A list of user rights (data) that informs a computer's operating system which permissions or access rights each user or group has to a specific system object, such as a directory or file. Each object has a unique security attribute that identifies which users have access to it, and the ACL is a list of each object and user-access privileges such as read, write, or execute.

Access Point (AP)   A centralized wireless device that controls the traffic in a wireless LAN. All traffic between the communicating computers must go through the access point.

ACL   see Access Control List.

Ad-hoc Wireless LAN   A Wireless LAN that consists of only computers with wireless LAN adapters.

AIEE   see American Institute of Electrical Engineers.

305

ALOHAnet One of the first wireless wide area networks. This wireless LAN consisted of seven computers that communicated in a bidirectional star topology that spanned four of the Hawaiian Islands, with the central computer based on Oahu Island.

American Institute of Electrical Engineers (AIEE) Formed on May 13, 1884, the American Institute of Electrical Engineers (AIEE) played an active role in the development of the Electrical Industry standards with primary focus on the wired communications, light, and power systems.

AP see Access Point.

ARPANET The precursor to the Internet, ARPANET was a large wide area network created by the United States Defense Department Advanced Research Project Agency (ARPA). Established in 1969, ARPANET served as a testbed for new networking technologies, linking many universities and research centers. The first two nodes that formed the ARPANET were UCLA and the Stanford Research Institute.

Asynchronous Transfer Mode (ATM) The mode of operation of the broadband integrated services digital network. All information in an ATM network that is to be transferred is first fragmented into small, fixed-sized frames known as cells. These are then sent over the network.

ATM see Asynchronous Transfer Mode.

Authentication The mechanism of ensuring that a rightful user is accessing the network by validating the authenticity of a user. The most common form of network authentication includes network and email logins.

Backoff Algorithm An algorithm that is used to calculate the duration to delay retransmission after a collision, before retransmitting in Ethernet.

Basic Service Set (BSS) When two or more wireless stations come together to communicate with each other, they form a basic service set. The minimum BSS consists of two stations.

Bluetooth A new short-range wireless communications standard that is used in handheld devices and mobile computers for limited data transfer and synchronization purposes.

Broadband A particular mode of operation of network data transmission that allows high data throughput. In a broadband operation, a number of separate data streams are simultaneously transmitted over a cable by assigning each stream a portion of the total available bandwidth.

BSS see Basic Service Set.

Bus Topology A network topology in widespread use for the interconnection of communities of digital devices distributed over a localized area. The transmission medium is normally a single cable to which all the devices are attached.

Carrier Sense Multiple Access with Collision Avoidance (CSMA/CA) A method similar to Carrier Sense Multiple Access with Collision Detection (CSMA/CD) used to reduce the collision between packets on a network that uses a shared medium by avoiding the collision of data in a shared medium.

Carrier Sense Multiple Access with Collision Detection (CSMA/CD) A method used to control access to a shared transmission medium, such as a coaxial cable, but to which a number of computers are connected. A station wishing to transmit a message first senses (listens to) the medium and transmits the message only if the medium is quiet—no carrier present. Then, as the message begins to be transmitted, the computer monitors the actual signal on the transmission medium. If this is different from the signal being transmitted, a collision is said to have occurred and been detected. The computer then ceases transmission and retries again later.

Carrier waves When electromagnetic waves are used to transmit data by superimposing the data on the radio waves, the waves carrying the data are known as carrier waves.

Caesar Cipher One of the oldest cryptographic algorithm that Julius Caesar used to send encrypted messages to his army. Caesar cipher is a substitution cipher. See substitution cipher for more information.

CF see Compact Flash.

Challenge Handshake Authentication Protocol (CHAP)   A network authentication protocol that mutually authenticates both the client and server using secret words that have been preinstalled in each system. In CHAP all user information including logins and passwords is transmitted in the network in encrypted form.

Challenge-and-Response–Based Authentication   A challenge–and–response–based authentication system provides a user to be authenticated with a challenge. For example, in dial-up networks, the server asks the dial-up user for username and password, and authenticates the user if the password provided by the user is correct.

CHAP   see Challenge Handshake Authentication Protocol.

Cleartext   Digital data that is transmitted without any encryption such that it can be analyzed without any processing is known as cleartext.

Compact Flash (CF)   An electronic circuit commonly used by handheld computing devices that is half the size of a credit card and adds computing features to the device using it. For example, a CF memory card can be used to add memory to a personal digital assistant (PDA).

CRC   see Cyclic-Redundancy-Check.

Cryptography   Derived from a Latin word cryptographia, cryptography means the enciphering and deciphering of messages in secret code or cipher. Today, cryptography is considered the art of protecting information by transforming it (encrypting it) into an unreadable format, called cipher text. Only those who possess a secret key can decipher (or decrypt) the message into plaintext.

CSMA/CA   see Carrier Sense Multiple Access with Collision Avoidance.

CSMA/CD   see Carrier Sense Multiple Access with Collision Detection.

Cyclic-Redundancy-Check (CRC)   A method used for the detection of errors when data is being transmitted. A CRC is a numeric value computed from the bits in the message to be transmitted. The computed value is appended to the tail of the message prior to transmission, and the receiver then detects the presence of errors in the received message by recomputing a new CRC and compares it with the CRC that is sent with the data.

Data Decryption   Decryption is the process by which an encrypted content is transformed to cleartext.

Data Encryption   Encryption is the process in which data in cleartext is transformed into an unrecognizable set of data characters for information security purposes.

Decipherment   see Data Decryption.

Network Gateway   A network device that routes network packets between networks.

Demilitarized Zone   Computers in a demilitarized zone are separated from the rest of the computers using firewalls and routers or sometimes physically to ensure that the data in the private LAN is not compromised.

Denial-of-Service (DoS)   A type of network attack in which an adversary makes the host computer so busy that it cannot reply to requests from genuine users.

DHCP   see Dynamic Host Configuration Protocol.

Dial-Up Connection   A type of network connection that is established between communicating entities by using modems over the phone line.

Digital Subscriber Line (DSL)   A type of broadband connection that provides high-speed connection to a private network or to the Internet.

Direct Sequence Spread Spectrum (DSSS)   A data transmission method for wireless networks in which the transmission signal is spread over an allowed band resulting in a transmission that is more resilient to wireless network jamming attacks.

Distribution System Service (DSS)   The services provided by a distribution service in a wireless LAN are known as distribution system service (DSS). A DSS provides five basic services: association, reassociation, disassociation, distribution, and integration.

Distribution System (DS)   A distribution system allows wireless LANs to be connected to the wired world.

DMZ   see Demilitarized Zone.

DNS   see Domain Name System.

Domain Name System (DNS)   The application protocol used in the TCP/IP suite to map the symbolic names used by humans into the equivalent fully-qualified network address.

DoS   see Denial-of-Service.

DS   see Distribution System.

DSL   see Digital Subscriber Line.

DSSS   see Direct Sequence Spread Spectrum.

Dynamic Host Configuration Protocol (DHCP)   A protocol for assigning dynamic IP addresses to devices on a network. With dynamic addressing, a device can have a different IP address every time it connects to the network.

EAP   see Extensible Authentication Protocol.

Encipherment   see Data Encryption.

Ethernet   The name of the local area network invented at the Xerox Corporation Palo Alto Research Center. It operates using the CSMA/CD medium access control method. The early specification was refined by a joint team from Digital Equipment Corporation (now Compaq), Intel Corporation, and Xerox Corporation and this, in turn, has now been superseded by the IEEE 802.3 standard.

Extensible Authentication Protocol (EAP)   A general protocol for authentication that also supports multiple authentication methods, such as token cards, Kerberos, one-time passwords, certificates, public key authentication, and smart cards.

Fast Fourier Transformation (FFT)   The Fast Fourier transform was discovered by a French mathematician and scientist, Charles Fourier, as a natural progression of his Fourier series theory. The Fourier series theory states that any waveform, however complicated, can be expressed as a series of two or more simple sine waves and cosine waves, if the waveform is periodical, that is, composed of the same repeated waveforms.

FCC ID   An identifier assigned by the United States Federal Communications Commission to devices that can emit radio frequency.

FFT   see Fast Fourier Transformation.

FHSS   see Frequency Hopping Spread Spectrum.

Firewall   A system designed to prevent unauthorized access to or from a private network. Firewalls can be implemented in both hardware and software, or a combination of both. Firewalls are frequently used to prevent unauthorized Internet users from accessing private networks connected to the Internet, especially intranets. All messages entering or leaving the intranet pass through the firewall, which examines each message and blocks those that do not meet the specified security criteria.

Frequency Bandwidth   The difference between the highest and lowest frequency signals that can be transmitted across a transmission line or through a network.

Frequency Hopping Spread Spectrum (FHSS)   A data transmission method for the wireless LANs in which the data carrier wave frequency oscillates within a band.

HTML   see HyperText Markup Language.

HTTP   see HyperText Transfer Protocol.

Hubs   A common connection point for devices in a network. Hubs are commonly used to connect segments of a LAN. A hub contains multiple ports. When a packet arrives at one port, it is copied to the other ports so that all segments of the LAN can see all packets.

HyperText Markup Language (HTML)   The authoring language used to create documents on the World Wide Web.

HyperText Transfer Protocol (HTTP)   A protocol used to send and receive data between a Web server and a browser.

IBSS   see Independent Basic Service Set.

IEEE   see Institute of Electrical and Electronics Engineers.

Independent Basic Service Set (IBSS)   A type of wireless LAN that does not contain an AP.

Industrial, Scientific, and Medical (ISM) Band   A frequency band reserved in most countries for industrial, scientific, and medical purposes. ISM bands generally do not require any licensing.

Infrared   An invisible electromagnetic radiation that has a longer wavelength than visible light.

Infrastructure Wireless LAN   A type of wireless LAN in which all data is transmitted through an access point.

Initialization Vector (IV)   A sequence of random bytes appended to the front of the plaintext before encryption by a block cipher. Adding the initialization vector to the beginning of the plaintext eliminates the possibility of having the initial cipher-text block the same for any two messages. For example, if messages always start with a common header (a letterhead or From line) their initial cipher-text would always be the same, assuming that the same cryptographic algorithm and symmetric key was used. Adding a random initialization vector eliminates this from happening.

Institute of Electrical and Electronics Engineers (IEEE)   An organization composed of engineers, scientists, and students. Founded in 1884 as the AIEE, the IEEE was formed in 1963 when AIEE merged with IRE, the IEEE is best known for developing standards for the computer and electronics industry.

Institute of Radio Engineers (IRE)   An engineering organization formed in the early 1900s as a result of merger between the Society of Wireless and Telegraph Engineers and the Wireless Institute, two separate organizations working on the wireless communication standards. IRE later merged with AIEE to form IEEE.

International Standards Organization (ISO)   An international organization composed of national standards bodies from over 75 countries. ISO has defined a number of important computer standards, the most significant of which is perhaps OSI (Open Systems Interconnection), a standardized architecture for designing networks.

International Standards Organization   Open Systems Interconnection (ISO/OSI) Reference Model. An ISO standard for worldwide communications that defines a networking framework for implementing protocols in seven layers. Control is passed from one layer to the next, starting at the application layer in one station, proceeding to the bottom layer, over the channel to the next station and back up the hierarchy.

Internet A Global collection of high-powered computers that are connected to each other with Network cables, telephone lines, Microwave dishes, satellites, and so on, to form a network.

Internet Protocol (IP) The network protocol used on the Internet for data communication. IP specifies the format of packets, also called datagrams, and the addressing scheme. Most networks combine IP with a higher-level protocol called Transmission Control Protocol (TCP), which establishes a virtual connection between a destination and a source.

Internet Protocol Address (IP Address) Every computer connected to the Internet must be assigned an IP (Internet Protocol) address. This address is a series of numbers such as 198.41.0.4 and acts much like a phone number. Whenever one computer wants to open a connection to another, such as when you want to connect to your mail server to collect your email, it first needs to know the IP address.

Internet Protocol Routing (IP Routing) The process of sending packets from one network to another through routers.

Internet Protocol Security (IPSec) IPSec is one of the most secure methods of setting up a Virtual Private Network (VPN). It allows you to join remote networks or computers so that they are effectively communicating directly without eavesdropping or tampering of data.

IP Address see Internet Protocol Address.

IP Routing see Internet Protocol Routing.

IPSec see Internet Protocol Security.

IRE see Institute of Radio Engineers.

ISM Band see Industrial, Scientific, and Medical (ISM) Band.

ISO see International Standard Organization.

ISO/OSI see International Standards.

IV see Initialization Vector.

Kerberos Kerberos is a freely available authentication protocol developed and invented by Massachusetts Institute of Technology (MIT) as a solution to network security problems. Strong cryptography is used in the Kerberos for both clients and server to prove their identities over insecure network connections.

LAN see Local Area Networks.

LDAP see Lightweight Directory Access Protocol.

Lightweight Directory Access Protocol (LDAP) A set of protocols for accessing information directories. LDAP is based on the standards contained within the X.500 standard, but is significantly simpler. And unlike X.500, LDAP supports TCP/IP, which is necessary for any type of Internet access.

Local Area Networks (LAN) A data communication network used to interconnect a community of digital devices distributed over a localized area. The devices may be office workstations, servers, PDAs, and so on.

MAC see Message Authentication Code.

MAC Address see Media Access Control Address.

MAC Authentication see Media Access Protocol Address Authentication.

Media Access Control (MAC) Address A hardware address that uniquely identifies each node of a network.

Media Access Protocol Address Authentication Some networks allow client authentication based on the MAC address of the network card. The MAC protocol address authentication scheme only authenticates the LAN card, not the actual user.

Message Authentication Code (MAC) A small chunk of data that represents the integrity and authenticity of a block of data. The MAC value is normally generated using a cryptographic method.

Microwave Microwaves are high frequency electromagnetic waves similar to the radio and TV signal waves and are used in radio and TV transmission, long distance telephone calls, and in radar. The microwave band starts from 915 MHz upwards.

Network Broadcast   A data transmission by a network node that is intended for all nodes on the network.

Network Interface Card (NIC)   A hardware device that is generally used to connect a computing device to a network. Most common types of NICs are Ethernet and the Token-ring NICs.

Network Traffic Based Attacks   A type of network attack in which a hacker or adversary modifies network data before it reaches an intended party. For example, if Alice sends some data to Bob, Eve the hacker may intercept the network traffic containing the data intended for Bob, modify it, and then send it to Bob.

OFDM   see Orthogonal Frequency Division Multiplexing.

Network Operational Security   A type of network security that is concerned with safeguarding, securing, and ensuring a flawless operation of a computer network. Network operational security assumes the roles of information assurance, personnel access control security (controlling who can access the network), defining authorization roles (restricts who can do what on a network), and physical security of the network equipment.

NIC   see Network Interface Card.

Orthogonal Frequency Division Multiplexing (OFDM)   A data modulation technique used for transmitting large amounts of digital data over a radio wave. OFDM works by splitting the radio signal into multiple smaller subsignals that are then transmitted simultaneously at different frequencies to the receiver. OFDM reduces the amount of cross talk in signal transmissions.

Packet Sniffing   An activity in which an individual uses a computer program to monitor and record all network communications.

PAN   see Personal Area Network.

PAP   see Password Authentication Protocol.

Password Authentication Protocol (PAP)   The most basic form of authentication, in which a user's name and password are transmitted over a network and compared to a table of name-password pairs. PAP is considered inherently weak because the passwords are transmitted to the server in cleartext.

Password Based Attacks   A network attack in which a hacker logs in to a computer using a password obtained by stealing or repeatedly submitting different passwords in an attempt to hit the correct access information.

PCI   see Peripheral Component Interconnect.

Peer-to-Peer Networks   A network in which different nodes directly talk to each other without requiring any central node. A wireless peer-to-peer network is one that does not use an access point.

Peripheral Component Interconnect (PCI)   A computer interface hardware design for enabling computer users to add functionality by using appropriate components that fit into the interface.

Personal Area Network (PAN)   A type of short-range networks that allows individuals to exchange data with a simple touch or grasp, such as a handshake.

Plaintext   see Cleartext.

Point-to-Point Tunneling Protocol (PPTP)   A network protocol that ensures security of data transferred over an insecure medium. PPTP is used in virtual private networks to provide data security over the Internet.

PPTP   see Point-to-Point Tunneling Protocol.

Print Server   A computer or a network device that handles printing requests from computers in a network. A print server is normally connected to one or more printers and to the network.

PRNG   see Pseudo Random Number Generator.

Pseudo Random Number Generator (PRNG)   An algorithm that generates a sequence of random numbers relative to the first number given. The first number that must be provided to a PRNG algorithm is known as a random number seed.

Radio Frequency (RF)   Any frequency within the electromagnetic spectrum associated with radio wave propagation is known as a radio frequency. When an RF current is supplied to an antenna, an electromagnetic field is created that then is able to propagate through space. Many wireless technologies are based on RF field propagation.

Radio Frequency Interference   An unwanted electromagnetic energy in the frequency range generally used for radio communications. For example, a buzzing noise that may occur in some audio and radio equipment when two stations may be transmitting signals at the same frequency.

RADIUS   see Remote Authentication Dial In User Service.

RC4   see Ron's Code 4.

Remote Authentication Dial-in User Service (RADIUS)   An authentication and accounting system used by ensuring a user's identity. For example, when you dial in to the ISP you must enter your username and password. This information is passed to a RADIUS server, which checks that the information is correct, and then authorizes access to the ISP system.

Repeaters   A network device used to regenerate or replicate a signal. Repeaters are used in transmission systems to regenerate analog or digital signals distorted by transmission loss. Analog repeaters frequently can only amplify the signal while digital repeaters can reconstruct a signal to near its original quality.

RF   see Radio Frequency.

Ring Topology   A type of local area network. All the devices are connected in the form of a ring and messages are transmitted by allowing them to circulate around the ring. A device can only transmit messages on the ring when it is in possession of a control token. A single token is passed from one device to another around the ring.

Ron's Code 4 (RC4)   An encryption algorithm invented by famous American mathematician Ronald Rivest.

Routers   A network device that connects one network to another.

Service Station Identifier (SSID)   The SSID, a 32-character unique identifier attached to the header of packets sent over a wireless LAN, differentiates one wireless LAN from another, so all access points and all devices attempting to connect to a specific WLAN must use the same SSID.

Shared Key Authentication   An network authentication scheme in which both client and server must possess the same key in order to mutually authenticate each other. In a smile shared key authentication, both client and server send each other some random data, and each party returns the data in encrypted form by encrypting the data, using the shared key. If the receiving party successfully decrypts the data using the shared key, the party assumes that the peer possesses the shared key.

Small Office Home Office (SoHo)   A small office or an office setting at a home. Generally, a SoHo is a business environment that employs fewer than ten persons.

SoHo   see Small Office Home Office.

Spread Spectrum   Spread spectrum is a form of wireless communications in which the frequency of the transmitted signal is deliberately varied. This results in a much greater bandwidth than the signal would have if its frequency were not varied.

SSID   see Service Station Identifier.

Star Topology   A type of network topology in which there is a central node that performs all switching and data routing functions.

Subnet   The name given in the ISO documents to refer to an individual network that forms part of a larger internetwork.

Substitution Cipher   An encryption algorithm in which each alphabet is substituted by another known alphabet. For example, assume that our original message was "APPLE." If we substitute all occurrences of letter A with letter K, P with Z, L with O, and E with T, the resulting message will be "KZZOT." To decrypt the message, we perform the reverse of this substitution.

Supplicant   The client computer in an 802.1X-based network that needs to be authenticated.

TCP/IP   see Transmission Control Protocol/Internet Protocol.

Time-Sharing System   A time-sharing system allows multiple programs to virtually execute at the same time by dividing the central processing unit time into slots and giving each program a slot of time in a cyclic manner.

Transmission Control Protocol/Internet Protocol (TCP/IP) The term used to refer to the complete suite of protocols including IP, TCP, and the associated application protocols.

Trojan Horse Viruses Trojan horse viruses are a common way for intruders to trick an authorized computer user into installing "backdoor" programs. These backdoors can allow intruders easy access to your computer without your knowledge, change your system configurations, or infect your computer with a computer virus.

Virtual Private Networks (VPN) A VPN is a private network of computers that uses the public Internet to connect some nodes. Because the Internet is essentially an open network, a security protocol, for example PPTP or IPSec, is used to ensure that messages transmitted from one VPN node to another are secure.

Virus Based Attacks A network attack in which an adversary uses computer viruses to degrade or destroy a network.

VPN see Virtual Private Networks.

Virtual Private Network (VPN) Gateway A computer or hardware device that manages a VPN connection between a user and a private network.

WAN see Wide Area Network.

War Driving A hacker activity where the primary purpose is to use the Internet services of other individuals and corporations. A war-driver generally roams neighborhoods, office parks, and industrial areas looking for unprotected networks and sometimes sharing this information on the Internet.

WECA see Wireless Ethernet Compatibility Alliance.

WEP see Wired Equivalent Privacy.

Wide Area Network (WAN) A general term used to describe any form of network, private or public, that covers a wide geographical area.

Wired Equivalent Privacy (WEP)   A security protocol for wireless local area networks (LANs) defined in the 802.11b standard. WEP is designed to provide the same level of security as that of a wired LAN. WEP provides security by encrypting data over radio waves so that it is protected as it is transmitted from one end point to another.

Wireless Ethernet Compatibility Alliance (WECA)   An organization made up of leading wireless equipment and software providers with the mission of guaranteeing interoperability of Wireless Fidelity (Wi-Fi) products and of promoting Wi-Fi as the global wireless LAN standard across all markets.

Wireless LAN Adapters   A network interface card that connects a computing device with a wireless LAN.

Wireless Roaming   A feature of wireless LAN that allows users of a wireless LAN to move about separate wireless LANs without losing a network connection.

References

1. R. Metcalf and D. Boggs, "Ethernet: Distributed Packet Switching for Local Computer Networks," *Communications of the ACM* 19(7): 395-403, July 1976.

2. Internet Security, Applications, Authentication and Cryptography (ISSAC), University of California, Berkeley, "Security of the WEP Algorithm," http://www.isaac.cs.berkeley.edu/isaac/wep-faq.html.

3. Wireless Ethernet Compatibility Alliance, "802.11b Wired Equivalent Privacy (WEP) Security," http://www.wi-fi.net/pdf/Wi-FiWEP Security.pdf. February 19, 2001.

4. S. Kent and R. Atkinson, "Security Architecture for the Internet Protocol," ftp://ftp.isi.edu/in-notes/rfc2401.txt. Copyright (c) The Internet Society, 1998.

5. S. Fluhrer, I. Mantin, and A. Shamir, *Weaknesses in the Key Scheduling Algorithm of RC4,* Eighth Annual Workshop on Selected Areas in Cryptography, August 2001.

Index

NUMERICS
2.4-GHz frequency band, 52
5-GHz band, 52
5-GHz migration products, 176
10Base2 specifications, coaxial cables, 25–26
10Base5 specifications, coaxial cables, 26
32-bit address space, 191
802.11 standards
  data security, 61–62
  frequency bandwidth, 59
  future of, 131
  operating modes, 62–63
  shortcomings, 68–69
  spread spectrum technology, 59
  SSID (service set identifiers), 123
  task groups, 57
  Web authentication, 128
  Web protocol weaknesses and strengths, 129–130
  wireless LANS, 51–52
  wireless local networking, 37
802.11a standards, 66–68
802.11b standards, 64–66
802.11g standards, 68–69
802.15 standards, 57–58
802.16 standards, 58
802.1X authentication protocol, 133–135, 226–227

A
access control
  authentication and, 88
  preventing, 88
  user groups, 90–92

access control lists (ACLs), 92
access points (APs)
  antenna diversity, 199–200
  box contents, verifying, 196
  with broadband connections, 48
  configuring, 200–202, 210
  defined, 47
  802.11 standards, 62–63
  example of, 194
  installing, best locations for, 148
  location, selecting, 197
  multiple AP configuration, 196
  non-overlapping AP configuration, 220–221
  overlapping AP configuration, 196, 218–220
  power, troubleshooting, 254
  product identification information, 197
  roaming, 63–64
  security requirements, 119–120
  single AP configuration, 195
acknowledgment (ACK) message, 47, 61
ACLs (access control lists), 92
adapters
  CF (Compact flash), 192
  installing, 202–205
  mishandling of, troubleshooting, 255
  PC Card, 191
  PCI (Peripheral Component Interconnect), 192
  software, installing and configuring, 206
  wireless LANs, 40–44
ad-hoc networks, 49, 222–223
Advanced Encryption Standard (AES), 106
American Institute of Electrical Engineers (AIEE), 56

323

Index

antennas
  diversity, troubleshooting, 254
  external, adding, 193
  multipath propagation, 199–200
antivirus software, 272
application-based attacks. *See* attacks
application layer, ISO/OSI model, 18–19
application program security, 102
APs. *See* access points
ARPANET, 5
asynchronous transfer mode (ATM), 7
attacks
  administration programs for, 97–98
  antivirus software, 272
  application program security, 102
  browsers and mobile code, 100
  brute-force, 95
  DoS (Denial of Service), 96–97
  email attachment-borne viruses, 98
  email forging, 99
  email spoofing, 99
  file server and disk space security, 101–102
  hidden file extensions, 100
  insertion, 263
  Internet chat programs, 99
  network appliance security, 102
  overview, 94
  packet-sniffing, 96
  password-based, 95–96
  security networks from, 100–101
  Trojan horse viruses, 97
  unauthenticated file-sharing, 99–100
auctions, purchasing LAN equipment at, 185
authentication
  access control and, 88
  challenge-response mechanism, 89
  802.1X authentication protocol, 133–135
  insecure mediums, 108
  link level, 121
  network adapter, 121
  network user, 89–90
  n-factor, 90
  open-system, 128
  password-based, 93
  process of, 90
  remote user, 93
  security policies and, 262
  shared-key, 128
  single factor, 89
  SSID-based, 122
  two-factor, 90
  username, 93
  See also security
authorization and privileges levels, 88

B

bandwidth
  congestion, troubleshooting, 258–259
  spread spectrum techniques, 43
basic service set (BSS), 49, 62
BBS (bulletin board systems), 7
blackouts, 87
Bluetooth, wireless standards, 53
broadband connection
  access points with, 48
  devices, obtaining, 154
brownouts, 87
browsers, attacks on, 100
brute-force attacks, 95
BSS (basic service set), 49, 62
bulletin board systems (BBS), 7
bus topology (networks), 14

C

cables
  coaxial, 25–26
  problems with, troubleshooting, 255
  twisted-pair, 24–25
  UTP (unshielded twisted pair), 25
Caesar cipher, 105
Carrier Sense Multiple Access with Collision Avoidance (CSMA/CA), 47
Carrier Sense Multiple Access with Collision Detection (CSMA/CD), 47
case studies, wireless LAN, 274–286
catalogs, purchasing LAN equipment from, 185–186
CERN (European Organization for Nuclear Research), 7
CERT Coordination Center (CERT/CC), 97
CF (Compact flash) adapters, 192
Challenge Handshake Authentication Protocol (CHAP), 109–110
challenge-response authentication mechanism, 89
channels, frequency channels, 214
chat programs, attacks in, 99
chips, defined, 44
ciphers, 104–106
Cisco Systems, 172–174
Class A addresses, 30
Class B addresses, 30
Class C addresses, 30
cleartext, 89, 103
clear-to-send (CTS) messages, 47, 61
client side configuration risks, 264
CNET-Shopper, purchasing LAN equipment at, 185
coaxial cables, 25–26

Index

commercial operators, wireless access and, 76–77
Compact flash (CF) adapters, 192
Compatible Time-Sharing System (CTSS), 4
configuring
  APs (access points), 200–202, 210
  RADIUS service, 232
connections
  broadband, access points with, 48
  Point-to-Point wireless requirements, 245
cryptographic tokens, 90
cryptography, 104
CSMA/CA (Carrier Sense Multiple Access with Collision Avoidance), 47
CSMA/CD (Carrier Sense Multiple Access with Collision Detection), 47
CTS (clear-to-send) messages, 47, 61
CTSS (Compatible Time-Sharing System), 4

D

DARPA (DOD Advanced Research Projects Agency), 28
database security, 102–103
data confidentiality and privacy, 138–139
Data Encryption Standard (DES), 106
data-link layer (ISO/OSI Reference Model), 22, 46–47
data security. *See* security
dead spots, locating, 148
DEC (Digital Equipment Corporation), 37
decryption, 104–106
Default gateway, 31
demilitarized zone (DMZ), 101
Denial of Service (DoS) attacks, 96–97
Department of Defense (DOD), 5
deployment, wireless LANs, 73–81
DES (Data Encryption Standard), 106
DHCP (Dynamic Host Configuration Protocol), 195, 228
Digital Equipment Corporation (DEC), 37
digital subscriber line (DSL), 96
direct-sequence spread spectrum (DSSS), 37, 44, 59–60
disaster recovery plans, 87
disk space, security, 101–102
distribution system services (DSSs), 50–51
DMZ (demilitarized zone), 101
DOD (Department of Defense), 5
DOD Advanced Research Projects Agency (DARPA), 28
DoS (Denial of Service) attacks, 96–97
DSL (digital subscriber line), 96
DSSS (direct-sequence spread spectrum), 37, 44, 59–60
DSSs (distribution system services), 50–51
dwell time, 44
Dynamic Host Configuration Protocol (DHCP), 195, 228

E

EAP (Extensible Authentication Protocol), 110
802.11 standards
  data security, 61–62
  frequency bandwidth, 59
  future of, 131
  operating modes, 62–63
  shortcomings, 68–69
  spread spectrum technology, 59
  SSID (service set identifiers), 123
  task groups, 57
  Web authentication, 128
  Web protocol weaknesses and strengths, 129–130
  wireless LANS, 51–52
  wireless local networking, 37
802.11a standards, 66–68
802.11b standards, 64–66
802.11g standards, 68–69
802.15 standards, 57–58
802.16 standards, 58
802.1X authentication protocol, 133–135, 226–227
electromagnetic waves, 40–42
Electronic Industry Association/Telecommunication Industry Association (EIA/TIA), 25
email attacks. *See* attacks
encryption, 61, 104–106
enterprise networks, 74–75
equipment, purchasing
  comparison tools, 184–185
  considerations, 183–186
  on Internet, 184–185
  mail-order catalogs, 185–186
  office supply stores, 186
  suggestions for, 186–187
  wired LAN Ethernet, 169–170
  wireless LAN, 169
equipment product code (EPC), 53
errors, network installations, 32–33
Ethernet, wired LAN equipment technologies, 169–170
European Organization for Nuclear Research (CERN), 7
extended service set (ESS), 49

326 Index

Extensible Authentication Protocol (EAP), 110
external network attacks. *See* attacks

F

Fast Ethernet LANs, 170
Fast Fourier Transformation (FFT), 46
Federal Communications Commission (FCC), 42
file extensions, hidden, 100
file security, 104
file sharing, unauthenticated, 99–100
firewalls, 154, 273
firmware, troubleshooting, 254
5-GHz band, 52
5-GHz migration products, 176
Frame Relay, 7
frequency-hopping spread spectrum (FHSS), 44, 60

G

General Electric (GE), 4
Gigabit Ethernet LANs, 170

H

hackers
  dictionary attacks, 95
  DoS (Denial of Service) attacks, 97
  overview, 85
hardware problems, troubleshooting, 254–256
health concerns, radio frequencies, 81
heat sources, AP location, 198
hereUare Communications Web site, 77
hidden file extensions, 100
HomeRF, 52
hubs, 26–27, 154
HyperText Transfer Protocol (HTTP), 18

I

IANA (Internet Assigned Numbers Authority), 29
IAPP (interaccess point protocol), 132
IBM (International Business Machines), 4
IEEE. *See* Institute of Electrical and Electronics Engineers
IFFT (Inverse Fast Fourier Transformation), 46
independent basic service set (IBSS), 63
Industrial, Scientific, and Medical (ISM) band, 54
Industry Standard Architecture (ISA), 38
infrared-based systems, 42

infrastructure mode, wireless LANs, 49–50
initialization vectors (IVs), 130
insertion attacks, 263
installations
  adapters, 202–205
  APs (access points), 199
  networks, 32–33
  ORiNOCO PC Card
    under Linux, 300–304
    under Mac OS, 296–300
    under Windows 98, ME and 2000, 287–293
    under Windows NT 4.0, 294–296
  VPN server, 239–240
Institute of Electrical and Electronics Engineers (IEEE)
  AIEE (American Institute of Electrical Engineers), 56
  802 wireless standards, 56–58
  history of, 56
  IRE (Institute of Radio Engineers), 56
  ISO/OSI Reference Model, views on, 23–24
  wireless standards, development of, 37
Institute of Radio Engineers (IRE), 56
interaccess point protocol (IAPP), 132
internal network attacks. *See* attacks
International Business Machines (IBM), 4
International Standards Organization (ISO), 17
Internet
  as computer network, 11–12
  connecting wireless LAN to, 216–218
  purchasing LAN equipment on, 184–185
Internet Assigned Numbers Authority (IANA), 29
Internet Protocol Security (IPSec), 115
Internet Relay Chat (IRC), 99
Internet Service Providers (ISPs), 11, 65
intrusion detection and containment, 272–273
Inverse Fast Fourier Transformation (IFFT), 46
IP addresses, 29–30, 257
IP routing, 31
IRE (Institute of Radio Engineers), 56
ISA (Industry Standard Architecture), 38
ISM (Industrial, Scientific, and Medical) band, 54
ISO/OSI Reference Model
  application layer, 18–19
  data Link layer, 22
  IEEE's view of, 23–24
  network layer, 21–22
  overview, 17–18
  physical layer, 23
  presentation layer, 19–20

Index

session layer, 20
transport layer, 21
for wireless LAN adapters, 41
ISPs (Internet Service Providers), 11, 65
IVs (initialization vectors), 130

K
Kerberos, 111

L
LAN/MAN (Local Area Network standards and Metropolitan Area Network standards), 52
LANs (local area networks), 3, 9
laptops, installing adapters in, 202–203
layers, ISO/OSI model, 18–23
link-level authentication, 121
Linksys products, 176–177
Linux, ORiNOCO PC Card installation, 300–304
local area networks (LANs), 3, 9
Local Area Network standards ad Metropolitan Area Network standards (LAN/MAN), 52
logic link control (LLC), 23
login
  authentication and access control, 88
  enabling for wireless LAN users, 232
  login-based security, 102

M
Mac OS, ORiNOCO PC Card installation, 296–300
mail-order catalogs, purchasing LAN equipment from, 185–186
Massachusetts Institute of Technology (MIT), 4
media access control (MAC), 23
megabits per second (Mbps), 7
messages
  ACK (acknowledgment), 47, 61
  CTS (clear-to-send), 47, 61
  Request/Identity, 134
  RTS (ready-to-send), 47
messaging system-based attacks. *See* attacks
Micro Instrumentation and Telemetry Systems (MITS), 6
MicrowareHouse, purchasing LAN equipment from, 186
microwave-based networks, 42
MIT (Massachusetts Institute of Technology), 4
mobile code, attacks in, 100

modulator demodulator (MODEM), 6
monitoring software, intrusion detection and containment, 272
multipath propagation, 199–200

N
National Science Foundation (NSF), 29
NetGear products, 178–180
network adapter authentication, 121
Network Address Translation (NAT), 229
Network Control Protocol (NCP), 28
network interface card (NIC), 17, 24
network layer (ISO/OSI Reference Model), 21–22
networks
  ad-hoc, 49, 222–223
  bus topology, 14
  cable and physical connections, 24–27
  Default gateway, 31
  development of, 4–8
  enterprise, 74–75
  installing, 32–33
  Internet, 11–12
  ISO/OSI Reference Model, 18–23
  LAN (local area network), 9
  NIC (network interface card), 17, 24
  PAN (personal area network), 11
  peer-to-peer, 8
  ring topology, 14–15
  software, 28
  star topology, 15
  subnet masks, 31
  topologies, selecting correct, 15–16
  VPN (virtual private network), 12–13
  WAN (wide area network), 9–10
Network Solutions Inc. (NSI), 29
network traffic-based attacks. *See* attacks
network user authentication, 89–90
n-factor authentication, 90
NIC (network interface card), 17, 24
Nordic Mobile Telephone (NMT), 37
NSF (National Science Foundation), 29
NSI (Network Solutions Inc.), 29

O
OFDM (orthogonal frequency division multiplexing), 46
office supply stores, purchasing LAN equipment from, 186
online stores, purchasing LAN equipment on, 184
open system authentication, 61, 128
operating systems (OS), 88

Index

operational security. *See* security
ORiNOCO
  PC Card, installing
    under Linux, 300–304
    under Mac OS, 296–300
    under Windows 98, ME, and 2000, 287–293
    under Windows NT 4.0, 294–296
  products, 174–176
orthogonal frequency division multiplexing (OFDM), 46
OS (operating systems), 88
overlapping AP configuration, 218–220

P

packet-sniffing attacks, 96
Palo Alto Research Center (PARC), 5
PAN (personal area network), 11
Password Authentication Protocol (PAP), 109–110
password-based attacks, 95–96
password-based authentication, 93
passwords
  cleartext, 89
  usage policies and, 139
PC Card adapters, 191
peer-to-peer networks, 8
Peripheral Component Interconnect (PCI), 152, 192
personal area network (PAN), 11
Personal Computer Memory Card International Association (PCMCIA), 38
personal digital assistants (PDAs)
  in conjunction with PCs, 11
  installing adapters in, 204
  WPANs (wireless personal area networks) and, 58
personal operating space (POS), 58
physical layer convergence procedure (PLCP), 23–24
physical layer (ISO/OSI Reference Model), 23
physical medium dependent (PMD), 23–24
physical network security. *See* security
physical security. *See* security
piggybacking, 22
PKI (public key infrastructure), 273
plaintext, 106
PLCP (physical layer convergence procedure), 23–24
PMD (physical medium dependent), 23–24
Point-to-Point Tunneling Protocol (PPTP), 114–115

POS (personal operating space), 58
presentation layer (ISO/OSI Reference Model), 19–20
print sharing, 145
privacy, 61–62, 262–263
privileges and authorization levels, 88
Pseudo Random Number Generator (PRNG), 62
public areas, wireless access in, 76–78
public key infrastructure (PKI), 273
purchasing LAN equipment. *See* equipment, purchasing

R

radio frequencies (RF)
  electromagnetic wave example, 41
  FCC regulations, 43
  health concerns, 81
  HomeRF, 52
  security requirements, 120–121
  spread spectrum modulation techniques, 43
ready-to-send (RTS) messages, 47, 61
Remote Authentication Dial-in User Service (RADIUS) Server
  configuring for wireless users, 232
  overview, 170
  vendors, 171
remote user authentication, 93
repeaters, 27
Request for Comments (RFC), 115
Request/Identity message, 134
resident-data security, 104
return of investment (ROI), 157–158
RF. *See* radio frequencies
RG Setup Utility, 211–213
ring topology (networks), 6, 14–15
roaming, 63–64
routers
  defined, 27
  intrusion detection and containment, 273
  IP routing, 31
  traffic flow between two networks through, 153
RTS (read-to-send) messages, 47, 61

S

SA (Station Adapters), 78
Secure Sockets Layer (SSL), 264
security
  application program, 102
  APs (access points), 119–120
  cryptography, 104

Index

database, 102–103
data confidentiality and privacy, 138–139
encryption, 61, 104–106
against external attacks, 100–101
file, 104
file server and disk space, 101–102
intrusion detection and containment, 272–273
network appliance, 102
network data, 115
physical network security, 87
privacy, 61
resident-data, 104
RFs (radio frequencies), 120–121
risks, 53–54
transmission, 106–109
UPS (uninterruptible power supply), 87
wireless LANs, 118–122
See also authentication
security policies
 communicating, 271
 compliance, 271
 creation guidelines, 265–266
 requirements for, 262–265
 sample of, 267–271
service set identifiers (SSID), 123
session layer (ISO/OSI Reference Model), 20
shared-key authentication, 61, 128
shared printers, 145
single factor authentication, 89
sites. *See* Web sites
site surveys, performing, 147–149
Small Office Home Office (SoHo), 65, 73–74
software
 antivirus, 272
 installation problems, troubleshooting, 257
 network, 28
spread spectrum
 DSSS (direct-sequence spread spectrum), 44
 802.11 standards, 59
 FHSS (frequency-hopping spread spectrum), 44
 network example, 43
 OFDM (orthogonal frequency division multiplexing), 46
SSID (service set identifiers), 123
SSL (Secure Sockets Layer), 264
standalone wireless LANs, 146, 215–216
star topology (networks), 15
Station Adapters (SA), 78
station services (SSs), 50
stations (STAs), 49

StoreRunner, purchasing LAN equipment at, 185
subnet masks, 31
surges, 87

T

task groups, 802.11 working group, 57
10Base2 specifications, coaxial cables, 25–26
10Base5 specifications, coaxial cables, 26
32-bit address space, 191
Time Share System (TSS), 5
T-Mobile Web site, 77
Token Ring, 6
topologies, network, 13–16
Transmission Control Protocol/Internet Protocol (TCP/IP), 4, 28–29
transmission security. *See* security
transport layer (ISO/OSI Reference Model), 21
Trojan horse viruses, 97
troubleshooting
 bandwidth congestion, 258–259
 hardware problems, 254, 256
 software problems, 256–257
TSS (Time Share System), 5
twisted pair cables, 24–25
two-factor authentication, 90
2.4-GHz frequency band, 52

U

uninterruptible power supply (UPS), 87
unshielded twisted pair (UTP) cables, 25
upgrades, wireless LANs, 259
username authentication, 93

V

value added resellers (VARs), 176
vendors
 Cisco Systems, 172–174
 Linksys, 176–177
 NetGear, 178–180
 ORiNOCO, 174–176
 RADIUS Server, 171
 Xircom, 180–183
ViperText Transfer Protocol (VTTP), 18
virtual private network (VPN)
 access by group member ship, 242
 access by user account, 241
 addresses and name servers, setting up, 241
 clients, configuring, 242–243
 connectivity, testing, 244
 installing and enabling, 239–240

330 Index

virtual private network (VPN) *(continued)*
  network transmission security, 108
  overview, 12–13
  problems with, troubleshooting, 257
  Windows 2000 VPN server, setting up, 238–239
virus-based attacks. *See* attacks
VPN. *See* virtual private network
VTTP (ViperText Transfer Protocol), 18
vulnerabilities. *See* attacks

W

wall mounts, 196
WAN (wide area network), 5, 9–10
war dialing, 131
war driving, 265
Wayport Web site, 77
WB (wireless bridges), 78
Web sites
  FCC, 533
  hereUare Communications, 77
  T-Mobile, 77
  Wayport, 77
  WiFinder, 78
  WLANA, 78
WECA (Wireless Ethernet Compatibility Alliance), 52, 54
WEP (Wired Equivalent Privacy), 52–54
wide area network (WAN), 5, 9–10
Wi-Fi (wireless fidelity), 52–53
WiFinder Web site, 78
Windows ME, ORiNOCO PC Card installation, 287–293
Windows 98, ORiNOCO PC Card installation, 287–293
Windows NT 4.0, ORiNOCO PC Card installation, 294–296
Windows 2000, ORiNOCO PC Card installation, 287–293
Wired Equivalent Privacy (WEP), 52–54
wired LANs, 169–170
wireless bridges (WB), 78
Wireless Ethernet Compatibility Alliance (WECA), 52, 54
wireless fidelity (Wi-Fi), 52–53
Wireless Internet Service Providers (WISPs), 65, 75–78
wireless LANs
  adapters, 40–44
  ad-hoc mode, 49
  benefits of, 72–73
  connecting to Internet using, 216–218
  costs associated with, 78–79
  coverage area and range of, 150
  data-link layer, 46–47
  data security, 122
  deployment issues, 79–81
  deployment scenarios, 73–77
  DSSs (distribution system services), 50–51
  801.11 standards, 37
  802.11 standards, 37
  equipment technologies, 169
  hardware problems, troubleshooting, 254, 256
  history of, 36–37
  infrared-based systems, 42
  infrastructure mode, 49–50
  login, enabling, 233
  microwave-based networks, 42
  OSI model for, 41
  overview, 35–36
  passwords and usage policies, 139
  planning process
    case study, 159–165
    hardware considerations, 151–154
    overview, 143–144
    potential users, communicating final plan to, 158
    requirements and expectations, setting up, 149–150
    ROI (return on investment), 157–158
    rollout, scope of, 147
    site survey, 147–149
    software considerations, 154
    wireless needs, understanding nature of, 144–147
  public access, 76–78
  security requirements, 118–122
  security risks, 53–54
  selecting correct, 145
  simple, 37–38
  standalone, 146, 215–216
  standards, 51–52
  supporting operating systems, 171
  upgrading, 259
wireless personal area networks (WPANs), 58
WISPs (Wireless Internet Service Providers), 65, 75–78
WLANA Web site, 78
World Wide Web (WWW), 7

X

Xircom products, 180–183